WORLDS IN SHADOW

Also available in the Bloomsbury Sigma series

WORLDS IN
SHADOW

*Submerged Lands in Science,
Memory and Myth*

Patrick Nunn

BLOOMSBURY SIGMA
LONDON · OXFORD · NEW YORK · NEW DELHI · SYDNEY

BLOOMSBURY SIGMA
Bloomsbury Publishing Plc
50 Bedford Square, London, WC1B 3DP, UK
29 Earlsfort Terrace, Dublin 2, Ireland

BLOOMSBURY, BLOOMSBURY SIGMA and the Bloomsbury Sigma logo are
trademarks of Bloomsbury Publishing Plc

First published in the United Kingdom in 2021

Photo credits (t = top, b = bottom, l = left, r = right, c = centre)

Colour section: P. 1: © Patrick Nunn (t); © Évariste-Vital Luminais (Wikimedia Commons /
Public Domain) (b). P. 2: © BBC (t); Dr Jon Henderson / Pavlopetri Underwater
Archaeology Project (c); © Patrick Nunn (bl, br). P. 3: © British Library Board G.6837
(tl, tr); © Theodore deBry (Wikimedia Commons / Public Domain) (b). P. 4: © Patrick
Nunn (t, c); © Lauren Juliff (b). P. 5: © Illustrated London News (t); © Patrick Nunn (c);
© David Lee (Wikimedia Commons / Public Domain) (bl); Williamborg (Wikimedia
Commons / Public Domain) (bl). P. 6: © Victor Ruiz (Wikimedia Commons / Public
Domain) (t); © Sémhur (Wikimedia Commons / Public Domain) (c); © Kevin Daniels (b).
P. 7: © Roland von Huene (t); © Patrick Nunn (cl, cr, b). P. 8: © Public Domain (t);
© Patrick Nunn (c); © Sémhur (Wikimedia Commons / Public Domain) (b).

A catalogue record for this book is available from the British Library

Library of Congress Cataloguing-in-Publication data has been applied for

ISBN: HB: 978-1-4729-8347-3; eBook: 978-1-4729-8349-7

2 4 6 8 10 9 7 5 3 1

Typeset by Deanta Global Publishing Services, Chennai, India
Printed and bound in Great Britain by CPI Group (UK) Ltd, Croydon CR0 4YY

Bloomsbury Sigma, Book Sixty-seven

MIX
Paper from
responsible sources
FSC® C020471

To find out more about our authors and books visit www.bloomsbury.com
and sign up for our newsletters

To Gerard Walshe and Shelagh Gordon –
one for the saddlebags as you head down the Elbe Valley

Contents

'Civilization exists by geological consent, subject to change without notice'

Will Durant, 'What is Civilization' (1946)

CHAPTER ONE

Introduction: Hearing the Past

It was 12 March 1947 and a blustery Wednesday afternoon in Skidegate off the west coast of mainland Canada. Marius Barbeau from the National Museum, distinguished anthropologist and folklorist just one year away from retirement, had not especially enjoyed the steamer journey from Victoria to the heartland of indigenous carving in the Haida Gwaii (also known as the Queen Charlotte Islands). But Barbeau, celebrated for his dogged quest to understand First Nations' cultures, was not easily distracted.

One of the main reasons for his visit to Skidegate was to collect Haida oral traditions using a new Ediphone Master Wax Voice Writer. One of his principal informants was 75-year-old Chief Gyitadzlius, more commonly remembered as Henry Young, 'a tall spare man … with a dry and dignified manner', quite a contrast to the 'short … stout' and undeniably vital Barbeau.[1] Young was an accomplished carver of totem poles and, being from a chiefly line, was steeped in his tribe's lore. He told Barbeau an extraordinary story.

Long ago, Young recalled, his people lived in north-west Haida Gwaii in a large village across from Frederick Island. One day, a group of children playing on the beach noticed a stranger some distance away, wearing a fur cape of a kind never before seen in Haida lands. Running up to her, one cheeky boy lifted the cape to expose the stranger's back, the sight of which made the children laugh and jeer. After the adults called their children away, the woman went to sit alone on the sand near the ocean's edge. The water rose to her feet, so she got up and moved a little distance up the beach. The water again reached her feet and so it went on until the ocean had climbed higher than ever before. It became clear to the Haida that their homes would shortly be flooded, so in panic they tied logs together to make rafts and, taking to the ocean, were able to save themselves. Young explained that because these crude rafts could not be steered, each drifted to a different place, a story that could be a distant memory of the time – thousands of years ago – when the first Haida peoples are known to have been dispersed by the rising of the ocean level here.

Young's story can be read as **myth**, especially the detail about the stranger and the unfamiliar fur cape she wore, as well as the implied power she had to raise the ocean level and punish the insensitive Haida by drowning the lands they occupied. But the story can also be read as **memory**, carefully passed on orally as part of Haida traditions for thousands of years, contributing to the history not only of modern Haida people but also of Haida lands and the changes these have undergone since they were first populated. But it is also **science**, a distant echo of ancient people's explanations of what happened to them. For this story is likely to recall a time when Haida people were affected and scattered by rising sea level, something that happened as much as 12,700 years ago here and on other ice-free islands west of the Canadian mainland.[2]

Like Young's narrative, ancient stories that may be memories of times when the sea surface was lower than it is today are found in other parts of the world. Some of the more numerous and compelling are from Australia where indigenous people have stories about times – at least 7,000 years ago – when the ocean surface was lower, when coastlines were further seawards and when what are now offshore islands were part of the mainland. Some stories from the Great Barrier Reef coasts, for example, may conceivably recall times when this great reef was all dry land and date from an incredible 13,310 years ago, requiring them to have been passed down coherently by word of mouth across more than 500 generations.

Another Australian example is Spencer Gulf, a sizeable isosceles-shaped inlet on the southern fringe of the continent that was dry land, occupied by people during the coldest time of the last ice age around 20,000 years ago. After the ice age ended and the ocean surface began to rise, Spencer Gulf gradually drowned. Local people still tell stories about this. These include the lagoons once being strung like a necklace of pearls along the axis of the Gulf, the types of creatures that could be found and hunted in particular places, even the tribal conflicts arising from competition for the rich resources found there. But all this came to an end after the sea level rose and flooded the area. The lagoons were submerged, the ecosystems disrupted and rival tribes forced to the sides of Spencer Gulf by the encroachment of the ocean. People's ways of life were irreversibly altered, yet their memories of how things had once been were kept alive – not just as part of a rich history defining the journeys through time taken by particular peoples but also, more pragmatically, to alert subsequent generations to what had happened and how people had survived the associated challenges that might occur again.

In Narungga (Aboriginal) stories recalling the inundation of Spencer Gulf, it is said this began when a giant kangaroo dragged a magic bone behind it as it walked along the axis of the Gulf, carving a channel into which the ocean poured. It is said 'the sea broke through, and came tumbling and rolling along in the track cut by the kangaroo bone. It flowed into the lagoons and marshes, which completely disappeared.' Aside from the existence and the role of a giant kangaroo, the details are all plausible, an effect of storm waves superimposed on rising post-glacial sea level off the lip of Spencer Gulf that shredded the dune barrier protecting the lowlands behind, allowing sea water in and drowning them. An event that modern science tells us probably occurred 9,330 or more years ago.

But do not dismiss the giant kangaroo too hastily for, as manifestly improbable as this detail may strike us today, it may well have represented the state of scientific explanation at the time these memorable events took place. Maybe there really was an enormous kangaroo, a physiological outlier, ascribed super powers. Maybe such a creature was imagined in the sky or under the sea. Maybe such a creature was part of the people's spiritual belief system at the time. Whatever it was, it is possible that it was blamed for the drowning of Spencer Gulf because people at the time determined this the most likely explanation for this life-changing event. And before we rush to deride such ideas, consider what many people do today when they are confronted by potentially life-altering events. They pray to deities that may be as real to them as a giant kangaroo was to the Aboriginal people of Spencer Gulf nearly 10,000 years ago.

The human need to identify a cause for a life-altering event is equally evident in the Haida story about the drowning of inhabited islands. The ancient Haida people simply could not fathom how their gods might permit something as terrible as the submergence of an entire island to happen. They needed to blame something for this aberration, so they attributed it to some children's unthinking teasing of a stranger, the mischievous lifting of her cape, and rationalised the events that followed as punishment for this. So far back in time did the drowning of this island and the dispersal of the Haida nation probably occur, that it is as challenging to explain the role of the stranger and her cape in ancient Haida world views as it is to understand the role of a giant kangaroo in millennia-old Narungga thought. But I would argue that we should not rush to dismiss these details as fanciful or discard them in the belief they are mere narrative embellishments until we can impartially assess whether they might have meaning.

The stories from Haida Gwaii and Aboriginal Australia neatly illustrate the three main sources of information from which we can today discover details about once-inhabited, now-underwater lands: science, memory and myth. Each can be complementary, meaning that when they are read correctly they may yield information that is unique. But, of course, if we are biased, even subconsciously, and demean or dismiss things like memory and myth because we do not know how to interrogate them, then we are likely to end up with an incomplete picture of the past. The purpose of this book is to try to rectify the situation, to demonstrate that each of these three information sources is potentially valid, something that gives a roundness to the past, a multi-dimensionality to history that personalises it and makes it more relevant to us today. For now, perhaps more than ever before, the past is relevant to the future. In a world where we are confronted by global change that is as contemptuous of human endeavour and individual aspiration as it is dismissive of political borders and agendas, understanding how our ancestors were affected by comparable changes and how they overcame these is at once a lesson in coping as well as a beacon of hope.

Worlds in Shadow

Some 20,000 years ago, the Earth was in the grip of an ice age. Vast tracts of now-inhabited land, mostly in the northern hemisphere, were buried under ice, much as the continent of Antarctica is today. To form these ice sheets, water had been sucked from the world ocean, the surface of which had dropped some 120m, almost 400ft, as a result. After the ice age ended and the world started warming a few millennia later, this land-grounded ice began melting, water pouring into the sea and raising its level within 10,000 years or so by the same 120m.

As a result of this sea level rise, entire landmasses disappeared, including many inhabited by people. Larger higher landmasses shrank, forcing people from their margins towards their interiors – or offshore. This loss of land, the loss of the opportunities it once represented and the histories it once embodied, have shaped human history to an extent that most of us underestimate.[1]

Consider if something similar happened today – see the map in Figure 2.1. It looks familiar yet strangely incomplete. It represents the conterminous United States as it is today except with a proportion of land removed equivalent to that lost as a result of sea level rise following the last ice age.[2]

This map measures the profundity of the land loss felt by people after the last ice age, especially in the period 7,000–15,000 years ago when most of this sea level rise occurred. Imagine how the lives of people in the United States might change if the same thing started happening today. What would be the implications of a 1,000km-wide gulf opening

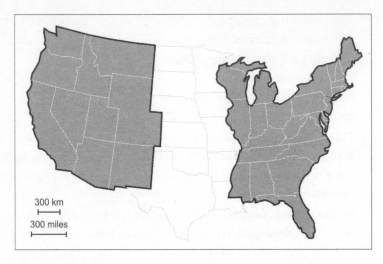

300 km

300 miles

Figure 2.1 *The conterminous United States with an area of land removed equivalent to that submerged by sea level rise after the last ice age.*

up between the eastern and the western parts of the country? Where would residents of the 11 affected states move to? Would people become anxious as rising seas began drowning the land on which they lived? Might there be conflict? Of course. All of these things surely happened at the end of the last ice age when rising seas drowned coastal lands across the planet.

Perhaps the least appreciated consequence of post-glacial coastal drowning is the erosion of human history, our history, a loss that inevitably renders the understanding of our contemporary situation incomplete. For most people on Earth today, history reaches only the coast of the lands they occupy; it does not stretch into times when that coast lay far from its present location, when the sea level was lower. For after the last ice age when sea level stopped rising, landmasses became essentially fixed in the minds of their human occupants, an empty yet bounded canvas on which society could develop. A place to civilise, freed from the threat of land loss. A place to tame.

The exciting prospect of uncovering within cultural traditions memories about actual lands, now underwater, is just part of this book, which presents a more complete account of submerged lands than most. Some of these lands we know about only because science has found them, reconstructed their forms and written their histories. But much of what science sometimes claims to have discovered was actually already known, preserved in cultural memories once dismissed as fantasies, but which we now recognise as preserving observations – eyewitness accounts if you like – of once-inhabited lands beneath the ocean. Observations that described the nature of these lands as well as the processes by which they became drowned.

This book focuses on lands that now lie beneath the ocean surface. Many writers refer to these lands as lost, vanished, disappeared, underlining the point that without a tangible connection to place, memories generally dissipate unless – as with the stories from the Haida and Aboriginal Australians recounted previously – there are special circumstances, especially cultural isolation and a pragmatic desire to pass on wisdom, that allow them to endure.

Over time the gradual loss of knowledge about drowned lands commonly leads to fragments of memory being incorporated into what many term folklore, a loose term for narrative expressions of apparent cultural inventiveness. Folklore has been studied mostly to gain insights into culture, but I would argue that it is also often history. For what is culture but the contemporary cumulative expression of people's pasts, whence they have come, what experiences and interactions they had along the way, and how they explained these?

We may not be able to readily detect an historical element in stories such as those about bigfoot, the yeti or the Australian bunyip, or indeed supposed folk tales like those in Greek or Norse cultures, but that does not mean that it was never there. This argument holds in other contexts besides stories. Some who study ancient Australian or South African rock art, for instance, have come to favour the idea that this was created for practical reasons, typically as memory aids, rather than being some indeterminate expression of creativity deep-rooted in human nature, which then evolved into modern art. We must be wary of superimposing our own beliefs and values on those of our distant ancestors who occupied quite different worlds and rationalised their existence in quite different ways from us.

The *conteurs* of north-west France – storytellers and custodians of Breton tradition – still travel as they have for thousands of years from town to town, village to village, even farm to farm, regaling eager audiences with ancient tales about the land. One of the most popular is that about the fabulous city of Ys, seat of King Gradlon, said to have been submerged by the ocean in Douarnenez Bay. No one today is quite sure where Ys was as its remains have never been found, but one story about why it was submerged is widely known. It concerns Dahut, the king's perfidious daughter, who one night opened the floodgates in the city walls at high tide, allowing the ocean in. Ys was flooded, abandoned, overwhelmed by the sea, and eventually slipped from history into the realm of myth.[3]

Or did it? Is the story of Ys – like that of Cantre'r Gwaelod in Wales (see Chapter 4) or Lyonesse off the coast of south-west England (see Chapter 8) – truly the myth most people suppose it must be? For if that is so, it seems a really strange subject for invention. Surely it is far more plausible to suppose (as science unequivocally shows) that the Atlantic coasts of north-west Europe, including Brittany, have been affected by a rise of sea level for thousands of years, a process that would unquestionably have drowned many inhabited places. Even perhaps one named Ys.

This is grounds for bringing stories about Ys out from the shadowy world of myth and back into the light of memory. We can regard them as distant echoes from a time long ago when a coastal community situated on the coast of Brittany was overwhelmed by the rising ocean, its surviving inhabitants forced to shift to less exposed places elsewhere. The memory of this tumultuous event stayed alive in the memories of the affected people, passed on to their descendants with increasing amounts of outrage and embellishment until the day arrived when the story came to be regarded as something less than factual. Perhaps 10 or 20 generations after all traces of Ys disappeared beneath the surface of the Atlantic, people started to regard the story of this as being so implausible that it could not be true. Yet this story remained cherished, told over and over again to ensuing generations, but without the burden of it being claimed as memory, just a good story that grew to be considered culture-defining.

While there are likely to be thousands of such stories – memories reclassified as myth – in all the world's longest-standing cultures, there are also instances of real myth masquerading as memory. Consider the nineteenth-century American seafarer Benjamin Morrell whose four lengthy voyages resulted in the apparent discovery of many new oceanic islands. With the benefit of scrupulous research, all these are now known to be mythical, invented. Morrell's motivation for inventing islands had to do with sealing, the main purpose of his voyages. His financial backers demanded not just full cargoes of fur-seal pelts and elephant-seal oil, but also information, concealed from their rivals, about islands where unexploited seal colonies might be found. Not really that keen on sealing it seems, Morrell realised the importance of reporting new islands (naturally overrun with seals) to sustain his backers' appetite for profit.

The first island he invented was in July 1825 in the north-west of the Hawaii group. He named it Morrell

Island and devoted three and a half pages of his logbook to a description of it, including the mouth-watering detail that its shore was 'lined with sea-elephants'.[4] The same month he claimed to have discovered Byers Island, likewise awash with 'seabirds, green turtles and sea-elephants'. In the South Atlantic Ocean, Morrell claimed to have located Saxenburgh Island – likely to have been an ocean mirage that deceived voyagers crossing the region a century earlier – and much more.

The likelihood that Morrell wrote the story of his four voyages only during the last of these and 'largely from a fertile memory' suggests that his wife Abigail, who often accompanied him, may have played some role in this. While little is known about her specifically, Mrs Morrell was dogged in her desire to sail with her husband, to the extent that she hid in his ship's bread locker when it left port on the third voyage so that the ship's owners, who had expressly forbidden her to sail with him, would not know she was on board. Whatever the true situation, Morrell's invention of places and his imaginative accounts of their economic potential place him among the least scrupulous of the voyagers in this age of nascent globalisation. Probably he was habitually deceitful, although some suggest he became mad after one of his ships was wrecked, while elsewhere his actions have been dismissed as those of an alarmingly incompetent navigator (Figure 2.2).

Having been created and fashioned in ocean-dominated worlds, the myths of Pacific Island peoples are quite different to those from other places. Take the stories explaining how islands formed. Some stories recall it was the larrikin demigod Maui, thrust on to the world stage in the 2016 Disney movie *Moana*, who repeatedly threw out

Figure 2.2 *Abigail and Benjamin Morrell: mad, bad or clueless?*

his magic fishing line to hook an island on the ocean floor and pull it to the surface. These islands did not always rise from their watery abodes without resistance. In some accounts, the fish-island thrashed violently as it was being pulled up; sometimes 'the waters rose bubbling and foaming ... and smoke came from [the depths] with a thunderous rumble and roar'. It is tempting to regard such details as embellishments to an original narrative, intended to enhance its memorability, to show how Maui had to strain to pull these huge islands up to the ocean surface, much as someone trying to land a large fish might do.

That interpretation is likely to be wrong, something given away by the bubbling and foaming and smoke and subterranean noises, all of which characterise shallow underwater volcanic eruptions. Often such eruptions lead to islands being formed, as we shall see in Chapter 12. And given that the ancestors of today's Pacific Island peoples traversed this vast ocean for 3,000 years or more, it is likely that they witnessed islands forming in this way on many occasions and rationalised their observations through stories about Maui, the legendary fisher of islands.[5]

Maui myths are also told about the Pacific island of Niue, a high limestone island showing no signs of a volcanic origin. Niue is actually a raised atoll, whose progress of emergence from the ocean is etched into its singular landscape. The central basin on the island is the former atoll lagoon, now 70m above where it formed. The rim around this basin is the ancient atoll reef, now spectacularly fossilised, below which lies a series of coral-reef terraces each representing periods of reef formation within the last 500,000 years. Ancient Pacific Islander voyagers, intimately familiar with coral reefs, would have instantly recognised high limestone islands like Niue as emergent, pieces of land raised from below the ocean surface. This is reflected in Niuean stories about the island's formation. One recalls a time when Maui lived here in a cave on the ocean floor, a time when 'the ocean rolled unbroken' over the site of the island. Then one day, flexing his mighty muscles, Maui pushed up the cave roof until Niue became 'a reef awash at low water', then with another great heave Maui 'sent it higher than the spray can reach'. And so, Niue was created.

Dismissing the Maui stories about Niue as myth misses the astonishing scientific insights they contain. For only in the past few decades have geologists understood how the landscapes of raised limestone islands like that of Niue must have developed – by the progressive, apparently staggered, upheaval of reef-fringed islands from beneath the ocean surface. Which is exactly what the Niuean stories tell us. The inescapable conclusion is that ancestral Niueans, 1,000 years ago or more, understood the geological origins of their island long before it fell under the gaze of modern science. Had scientists only known, they could have built on these ancient insights rather than going to the trouble of deducing them anew.[6]

Of course, myths may not always be grounded in fact. Throughout the ages, across thousands of years, storytellers keen to engage their audiences and retain their attention have embellished the bare bones of their narratives. The challenge for people today seeking to analyse potentially meaningful myths is to peel away the layers of embellishment to expose an empirical core. This is not always a straight-forward task; indeed some myths may not have such a core. Like a modern work of fiction, they may be wholly imaginary, something that can sometimes be exposed through knowledge about their creators and their motivations.

Take Atlantis. The name resonates across time and place. For many people, it is pregnant with hidden meaning. A key to the great mysteries about the history of the planet Earth, the origins of its peoples, and perhaps the proof that in the dim and distant past our ancestors had advanced well beyond what is generally believed. But for all the speculation, Atlantis never actually existed.

There is no escaping this. There are numerous clues in the writings of Plato, who manufactured the story of Atlantis in about 350 BC, that it is allegorical not factual. And there is simply no possible means by which anything the size of Atlantis, claimed by Plato to be around 240,000km^2 in area, could abruptly and violently have sunk beneath the ocean surface as he described.[7] So why do so many people insist otherwise? Why are they convinced that Atlantis was a real place? The short answer is because they want to believe it. Nothing more. Many people seek mystery to enliven their existence; their minds are soothed by imagining places like Atlantis, which are mysterious, malleable and thus full of promise; their details have not been soured by scientific revelation.

While Plato is today revered for a range of philosophical insights, it was in two of his later works – *Timaeus* and *Critias*, written when he was in his 70s – that 'a very strange story' is recounted. *Timaeus* is a book of largely cosmological speculations, as part of which the character Critias tells how his great-grandfather Solon, while travelling through Egypt between 593–583 BC, discussed ancient history with a group of priests. In recounting the ancient history of Athens, one priest told Solon that around 9600 BC there was a great Athenian empire rivalled only by that on an island west of the Pillars of Heracles (the modern Strait of Gibraltar) named Atlantis, which was the centre of a great empire. The island was larger 'than Libya and Asia taken together' and was surrounded by smaller islands. At one time, the Atlanteans extended their empire well into the Mediterranean, but were finally defeated by the Athenians. Later there were 'violent earthquakes and floods' that affected both Athens and Atlantis, the latter disappearing in one day and night 'beneath the sea', making the Strait of Gibraltar impassable to ships thereafter.

One of the reasons Plato cited for the success of the state of Atlantis was the way it was organised and run, precisely along the lines proposed by him in perhaps his most famous utopian work, *The Republic*. This remains key to understanding why both Atlantis and the contemporary Athens are fabrications. Plato intended them to illustrate the practicability of the systems of government and social organisation proposed in *The Republic*, which he had unsuccessfully lobbied the rulers of the city-states of Athens and Syracuse to adopt.

More details about Atlantis appear in Plato's unfinished *Critias*. Like other stories about fictional lands, the civilisation centred on Atlantis was advanced compared with its neighbours. From its beginnings, when it was the fiefdom of Poseidon, the Greek god of both the ocean and of earthquakes, Atlantis was well endowed with natural resources. Poseidon surrounded the hill (where his mistress resided) in the centre of the circular island with concentric rings of water and land. But as time wore on and the bloodline of Poseidon became ever more diluted, moral decline spread within Atlantis. The last part of *Critias* describes how the people of Atlantis changed from 'law-abiding, gentle and virtuous' citizens to being in a state of 'ugliness, unhappiness, and uncurbed ambition'. *Critias* ends abruptly as Zeus is about to address the assembled gods on the subject of punishing Atlantis.

Plato's *Timaeus* was intended as a sequel to *The Republic* and it is clear that the story of Atlantis in *Timaeus* and *Critias* is allegory, a fiction created to illustrate the principles explained in *The Republic*.[8] There are people who will read all this and still insist that Atlantis was a real place, just one that Plato found to be a convenient vehicle for his messages about politics and society. Yet had a wondrously advanced civilisation of the kind Plato described really existed 12,000 years ago or more, it would have certainly been recorded in many other sources besides that of a single book, written nearly 9,000 years later by a writer renowned for literary artifice. Had a massive island-continent like Atlantis disappeared beneath the ocean, abruptly or not, the traces of it would have been found long ago by geologists, but they have not.[9] Today we have good maps of every part of the ocean floor; Atlantis is simply not there.

Yet Plato saw nothing wrong in incorporating details of actual events into his Atlantis story, especially its dramatic end, to enhance its believability and thus its memorability. Far easier, especially in the oratorial academy of which Plato was part, to remember stories involving divinely instigated catastrophe than absorb dry details of competing models of statehood. Some of the events likely to have influenced Plato's stories of Atlantis include the island-destroying eruption of Stronghyle-Santorini about 1600 BC (described in Chapter 12) and the massive earthquake in 464 BC, 37 years before Plato was born, which flattened the city of Sparta, reconfiguring the power balance in this region almost overnight.[10]

This book takes up the challenge of finding a way through the myriad accounts of vanished lands recognisable through science, memory and myth to present a picture, shorn of embellishment, of this astonishing and poorly covered topic. It is key to the understanding of whom we, modern humans, actually are and where we have come from. It is also relevant to the future because today we live not only in a drowned world, but also a drowning world where the ocean surface is currently rising faster than at almost any time within the past 10,000 years or so. The outlook is confronting yet far from hopeless, but as the foremost rational species on the planet we have a duty of care that should be informed by fact and deduction rather than prejudice and guesswork.

On these pages you will find stories about lands submerged comparatively recently in human history, separated into those whose former existence is undisputed, those whose existence can be doubted and those which probably never existed. You will also find much older accounts of submerged lands, some having been occupied by our ancestors and others considerably predating the appearance of modern humans on Earth 200,000 years ago.

I will also look at why lands vanish and how they become drowned; then the effects of ocean-surface (sea level) changes, land-level changes driven mostly by crustal shifts, gravity-driven collapses, giant waves and the effects of volcanic eruptions.

Bringing us up to today, I examine where and why land may disappear in the future, what we have been doing about this, and how we might come to terms with expected future land loss.

CHAPTER THREE

Recently Drowned Lands

There are few images more poignant than that of a land, once inhabited by people like us, which has disappeared. Its disappearance takes with it a profound and meaningful history of associations between the living and the terrestrial, often lamented, and drops it into the featureless ocean of oblivion. No more will people be able to touch what once was touched, to feel what once was felt, or to understand what it really meant to inhabit such a place.

In 1608, John Smith, leader of the newly founded Jamestown Colony on the east coast of what became the United States, set out to explore massive Chesapeake Bay. Bad weather forced him ashore on the islands in its centre. He was not happy. 'Two dayes we were inforced to inhabite these uninhabited Isles; which for the extremitie of gusts, thunder, raine, stormes, and ill wether we called Limbo.'[1] One island visited by Smith is what is now known as Tangier Island, a place he found covered with trees. By 1939, more than three centuries later, Tangier was treeless although some of its older residents could remember a small grove of pine trees before it became submerged.[2] Today, as shown in Figure 3.1, the ocean has driven across centuries-old graveyards on Tangier Island, scattering gravestones and the bones of the dead across the shore, something that has shocked locals. 'I thought we could do better than this as a people. We can't abandon our towns, our graveyards, to the water like that,' said marine scientist David Schulte.[3]

This situation is playing out today along numerous other parts of the Chesapeake Bay coast. Born and raised on

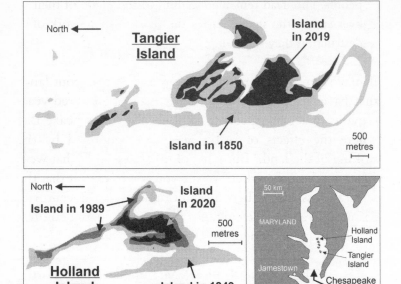

Figure 3.1 *Disappearing islands of Chesapeake Bay, eastern United States.*

Hoopers Island, businessman Johnny Shockley recalls with nostalgia the woods of his boyhood, noting that today, 'Those trees are dying. It's obvious. Where the marshes meet the woods, the marsh is taking over.'[4] Nearby Holland Island has shrunk phenomenally since it was first inhabited, as the map in Figure 3.1 shows. Even at its apogee around the start of the last century when it sustained a population of more than 360 (and had its own baseball team), Holland Island was getting smaller, its periphery being steadily eaten away through erosion attributable to rising sea level. Residents started leaving by 1914 and the island was uninhabited by 1922. Its obituary was penned that year.

> The paradise of the Chesapeake is no more. Today at low tide only vestiges are seen and at high tide the island is completely under water. Gone are the homes of the

people. The dead remained faithful to the island of their lives and as the storms drove the tides to their work of destruction they slept on.[5]

It may seem no surprise that the sense of loss about lands that have been submerged in the past few hundred years remains so acute, especially in places like Chesapeake Bay where the effects of rising ocean are amplified by the sinking of the land.[6] But what of inhabited lands that were submerged much earlier? While you might expect that the sense of loss would become more muted as the centuries passed, there is ample evidence from many coastal cultures that it can persist longer than you might expect.

Half a world away from the east coast of the United States, in the balmy islands of Micronesia in the north-western Pacific Ocean, we find a group of islands, rarely visited by outsiders, named Yap. From its unique language, the roots of which continue to puzzle linguists, to the relaxed approach to life – no Yapese driver ever exceeds the 25-miles-per-hour speed limit – that envelops the inquisitive visitor from the moment of their arrival, Yap appears the consummate enigma.

One of the Yapese islands once disappeared. It was named Sipin and was reportedly home to some of Yap's more traditional people, steeped in the ways of their ancestors, resisting the encroachment of the new. The story goes that one day the chief of Sipin decided things in this regard were becoming intolerable so he caused the island to disappear under the sea. At first, although it was wholly underwater, Sipin was still visible to people sailing through the area. Yapese stories tell how on a quiet night the torches of people moving around the sunken island could be seen below the water, their voices clearly audible.[7] And before you wonder how this might possibly be, consider a couple of other examples.

Crossing to the South Pacific, there are similar stories from the islands of Vanuatu. Despite being one of the world's more disaster-exposed nations – there are many active volcanoes and frequent earthquakes, as well as the ever-present threat of tropical cyclones (hurricanes) in summer – the people of Vanuatu evolved a culturally negotiated relationship with disaster that led them to regard it through a somewhat different lens to that which might appear self-evident.

There are three parallel lines of islands in Vanuatu, the central one being where most active volcanoes, above and below the ocean surface, are found today. These include the island of Ambae, whose inhabitants have recently been evacuated several times when the threat of explosive eruption appeared imminent; and the island of Kuwae, which blew itself to pieces in the year 1453 in what volcanologists regard as one of the largest eruptions anywhere on earth within the past 10,000 years (see Chapter 12).

Like the islands of Yap, those of Vanuatu adjoin a boundary between two giant chunks – known as plates – of the Earth's crust. The volcanism and earthquakes that affect Vanuatu arise from the unremitting thrusting of the massive Indo-Australian crustal plate beneath its Pacific counterpart. As the two plates grind past each other in a characteristic stick-slip, stop-start fashion many kilometres beneath the land surface, so earthquakes rattle the islands every time a slip occurs. Likewise, when the down-going plate reaches depths of several hundred kilometres or more, it starts melting, the liquid rock it forms rising upwards, sometimes reaching the ground surface where it pours out of volcanoes.

The combination of steep, undersea slopes and massive earthquakes explains the multitude of stories about vanished islands that exist in the traditions of the Ni-Vanuatu, the name for these islands' people.[8] One story recalls when two inhabited islands – Malveveng and Tolamp – abruptly sank. Located off the east coast of Malekula Island, these islands rose from the steep undersea slopes that plunge 2km to the floor of the ocean basin separating Malekula from the central chain of active volcanoes. One day, it seems likely, a massive earthquake shook this part of Vanuatu, triggering an equally massive undersea landslide that carried Malveveng and Tolamp downslope, causing their above-sea parts where people were living to be submerged.

There is no mystery about where these islands were. Today, Malveveng and Tolamp are shoals, prized fishing spots for people from nearby islands, respectively 5m and 15m below the ocean surface. It is said that when the sea is clear, divers at both sites can look down and see ancient *nasara* (meeting places), slab-paved roads, stone fences, *tamtam* (gongs) and sacred stones used in *namanggi* (ceremonies). This is one way in which local residents keep their memories of these former islands alive. And as we shall see in Chapter 9, there are also stories, dances and ceremonies that recall what happened here and why.

Another example, published here for the first time, comes from the southern Fiji Islands, east of Vanuatu. It concerns the island Solo (meaning 'rock' in the local dialect), where today stands a prominent lighthouse. Sitting on its steps, one looks out across a broad shallow lagoon, the position of its coral-reef-fringed margin detectable only from the sight of spray from distant crashing waves. There is not much room on Solo today for anything except the lighthouse, but this apparently was not always the case.

Numerous stories tell that Solo was once far more extensive – a place called Lomanikoro, which was home to an influential group of people. One day, the island shook and most of it was quickly submerged, its surviving inhabitants forced to move elsewhere. This not only impacted the geography of southern Fiji, but memorably reconfigured the power relations of its inhabitants. It is said that today people crossing the site of the drowned land can see the *yavu* (house mounds) below, and hear cockerels crowing and mosquitoes buzzing beneath the ocean surface.

It is an eerie experience to sail to Solo. You have the sense of being temporarily under the control of unknown forces, a sense heightened by the rituals by which local people abide when travelling there. As your boat enters the Solo lagoon through the reef passage, your speed must slow, everyone in the boat should refrain from talking loudly, should remove their headgear and then *tama* (squat and clap in respect, as when approaching a high-ranking person in Fiji). The instruction is to take only the fish you need – any more than that and your boat will never return home afloat.[9]

While these stories from islands in the western Pacific may seem far removed from those about land loss in Chesapeake Bay, in reality they are not. For while the loss of place in the Chesapeake is mourned in ways with which many readers of this book may readily identify, land loss is remembered and regretted in the Pacific Islands in ways you would expect in non-Western cultures where traditional knowledge has been conveyed orally along the river of history.

Yet such stories are not unique to the Pacific Islands. We find comparable examples in north-west Europe. Old tales from the Isle of Man off the north-west coast of England, for instance, talk of the existence of a sunken land and report that 'sailors assert that they frequently hear cattle lowing, dogs barking, beneath the waves'. On the nearby English mainland, a few kilometres from the centre of Blackpool, where the village of Singleton Thorpe disappeared in about 1554, the remains of the road leading to its submerged inn were still visible a century ago and fisherfolk feared the laughter from the inn's ghostly revellers that could sometimes be heard coming from beneath the waters.[10]

During the last great ice age – and most of the 20 or more preceding it – ice sheets covered much of the land in higher latitudes, especially in the northern hemisphere. Driven by changes in temperature and precipitation, great ice tongues called glaciers alternately pushed out or drew back along the fringes of these ice sheets. Imagine what happens when an ice tongue pushes out. It drives across the ground like a massive bulldozer pushing all the rocks and soil it encounters either off to its sides or in front of it. When that same ice tongue melts and withdraws, great piles and ridges of bulldozed rock are left behind. These are called 'moraines' and are prominent features of modern landscapes along the margins of former ice sheets. In the north of England, in the East Riding of Yorkshire, there existed during the last ice age a sizeable coastal inlet that became filled with moraines, forming a lowland landscape known as Holderness.

For scientists interested in why and how coastal erosion occurs, the soft unresisting morainic cliffs of Holderness have become a classic study site. The cliffs, which reach almost 40m in height, have been cut back in places by as much as 4m every year in recent times – and there is a long history of constructing sea defences to protect the valuable farmland behind them.

There is also a lengthy recorded history of land loss around Holderness, both along its ocean-facing eastern side and its southern side, which runs along the estuary of the River Humber. It was here, near the modern village of Skeffling, that in June 1219 a group of French monks was given the chapel of Burstall as a base from which to spread

Catholicism throughout this part of 'pagan' England. This was part of the wave of proselytisation that swept the country in the wake of the Norman Conquest in 1066–72, after which William the Conqueror divvied up the land among his most loyal lieutenants. The Albemarle family was thus rewarded with the Holderness area and brought monks from their Normandy home to occupy chapels like that of Burstall, later extended and known as Burstall Priory.

The spread of Catholicism in this part of England came to an end in 1536–41 when King Henry VIII dissolved the monasteries and founded the Church of England. While the history of Burstall Priory up until the Dissolution is well known, its subsequent history – detached from church records – is less so. It was in ruins by 1721, when the sketch in Figure 3.2 was made, and closer by a kilometre or more to the shoreline of the Humber Estuary than at the time of its establishment. Within 100 years, Burstall Priory had disappeared, its former site today indiscernible among the tidal flats.[11]

Figure 3.2 *The ruins of Burstall Priory on the shore of the Humber Estuary in the East Riding of Yorkshire, England, in 1721.*

South from Holderness, one reaches East Anglia, another moraine-choked part of eastern England where coastal change has also been conspicuous over the past thousand years or so. Located on the Suffolk coast adjoining the North Sea, Dunwich is one of a string of coastal villages here, prominent in medieval trading networks and once far larger, but which coastal erosion reduced in size faster than this prominence could be sustained, eventually forcing its merchants to abandon it for less-exposed towns like Ipswich at the heads of nearby estuaries. For the past few hundred years, Dunwich has been physically eroded, with many buildings including a number of iconic churches toppling into the sea at the foot of the retreating cliffs.

The Dunwich coast comprises loose, readily eroded rocks, both those accumulated at the former river mouth, and their older bedrock counterparts made from poorly cemented sands and gravels washed out from ice-bulldozed moraines. Waves come at Dunwich from almost every direction except west, the highest from the north and north-east, slicing into the cliffs and causing their collapse and retreat. Here this process has been amplified because this part of the English coast is also sinking, a result of the large-scale warping of the Earth's crust in the southern North Sea basin, and today is also affected by rising sea level. The net effect of this – sinking plus sea level rise – is to expose Dunwich's fragile shores to faster erosion than is occurring along many other North Sea coasts.

The sun casts a long shadow across the Dunwich coastline in the late afternoon, rendering the sea dark and fathomless. Lights come on in the pretty, quintessentially English village, speckling the marsh mists drifting in from the north that might otherwise envelop it. At such times older residents often swear they can hear the mournful sounds of tolling bells, not bells from any churches on the land but those beneath the sea, rung perhaps by ghosts, perhaps by waves churning around the drowned church steeples.

Stories of tolling bells from submerged churches are common elements of folklore elsewhere along north-west European coasts, but are almost certainly all apocryphal. At least in the case of Dunwich, all the valuables (including bells) were stripped from the churches long before they fell into the sea. And why would we suppose otherwise? Yet stories of tolling subterranean bells serve the purpose of keeping alive memories of ancient places, in this case those that vanished beneath the ocean surface. Perhaps these stories could have been kept alive by being written down, but in places like medieval Dunwich, where most residents were not literate, written accounts might have barely impacted people's recollections. So oral traditions, passed on from one generation to the next, would have been a more powerful way of retaining these memories in popular culture.

Such details also delay the inevitable slide of memories of land loss into the shadowy realms of legend and myth. In places like Dunwich, it once helped give younger residents a sense of identity and attachment, to appreciate the losses their predecessors endured; it gave them a sense of their place in history. In this way, stories of eerie bells tolling off the Dunwich coast are not so different from Fijian tales about the buzzing of mosquitoes irritating the people on the sunken island in Solo lagoon, or Yapese stories about hearing people conversing underwater on the submerged island of Sipin. In distinctive cultural contexts, memories of past events are kept alive using familiar imagery.

It is easy to understand why some ancient cities were built close to the sea. Ever the driver of economic prosperity, most trade before the nineteenth-century Industrial Revolution was carried out across water, interaction being generally easier and more dependable than that across the land. In many parts of medieval Europe, for instance, busy seaways and rivers saw goods moved from the places they were produced to places they were in demand.

River-mouth cities were also places in which power from the accumulation of wealth became concentrated. The wealth was protected by harbour-entrance defences, the power secured by fleets of warships. For example, in thirteenth-century Venice, the hugely successful trade nexus near the apex of the Adriatic Sea, a fleet controlled access to the port and deterred aggressors, sometimes far from home. The establishment of cities in river-mouth and delta-edge locations was also a feature of European colonial occupation in more recent times. Cities like New Orleans (United States), Penang (Malaysia) and Port Harcourt (Nigeria) began this way.

Yet coastal-lowland and river-mouth locations sink and at times – like today – when sea level is rising they are regularly flooded, especially during storms, so we find that many such cities have a long history of combatting the ocean. Some have won, at least temporarily, by building walls. Some were defeated, becoming submerged, sometimes lost from history.

Submergence of coastal towns and cities is likely to be known by historians of the distant future as a widespread twenty-first-century phenomenon. But it is not one lacking earlier precedents, as illustrated by the example of Vineta, located somewhere just inland of the Baltic Sea margin of northern Europe. In the early middle ages, from the tenth to twelfth centuries, Vineta was acclaimed as the most important and powerful port city in northern Europe. Writing about AD 970, the peripatetic Ibrahim ibn Yaqub described it as 'a large city by the ocean with twelve gates, the greatest of all cities in Europe [located] in the country of Misiko in the marshes by the ocean'.[12]

Once home to between 8,000 and 10,000 people, Vineta is no more.[13] In truth, there is disagreement as to exactly where it was located, so changeable has been the environment of the area. Most researchers contend that Vineta was located close to modern Wolin in Poland although some place it across the German border.

The disappearance of Vineta has popularly been ascribed to the bad behaviour of its inhabitants when the city was in its heyday, but the reality is likely more prosaic. Vineta was built across a marshy island in a river delta, probably on a bank of the Dziwna River, one of the three distributaries of the Oder River. A thousand years ago, this was a prime location for what the city founders envisioned. The Oder was an important thoroughfare, a conduit for trade between Scandinavia and the Baltic states on the one hand and central Europe on the other.

Let us assume for the sake of argument that Vineta was synonymous with the medieval port town of Wolin, whose history is better documented. In the tenth century, Wolin-Vineta was described as:

> A cosmopolitan and boisterous place … [where] one might have seen Arab, Byzantine, Jew, Saxon, Pole, Russian, Scandinavian and Wend meeting at the town's lighthouse, walking down Wolin's oak-paved streets, or standing before its temple to the deity Svantevit. In one of its markets, one might rarely glimpse silk from China or jewellery from Syria …

Yet within 200 years, its decline was almost complete. It had become 'a wasted shell of its once-grand self'.[14]

It has been proposed that a series of storm surges led to the abandonment of Wolin-Vineta, giant waves driving through the Oder Delta pulverising everything in their path. But such places are exposed to multiple sources of environmental stress. The channels of delta distributaries are notoriously mobile because of the huge volumes of water and sediment they are periodically forced to accommodate, so it is equally likely that a flood associated with prolonged heavy rainfall may have caused the Dziwna channel to shift sideways, literally cutting away the soft foundations of Wolin-Vineta. Another option is avulsion – the abrupt abandonment of one river channel for another – that can also be triggered by large floods in deltas. An alternative is that the people of Wolin-Vineta overused their natural resources (particularly trees and soil) so completely that they were eventually forced to import food, leading to a situation in the mid-eleventh century in which the town 'became too exhausted to repulse, or recover from, sea-borne raids'.

Whatever the exact reasons for Vineta's disappearance, it was no more by the sixteenth century, even though its existence was evidently sufficiently well recollected at the time for its former location to be marked, tantalisingly vaguely, on a contemporary map.

Centuries earlier in Egypt, around the mouths of the River Nile, something similar happened. Given that it is by far the longest river in Africa, the Nile has a massive river delta at its mouth which opens into the eastern Mediterranean. Like many river deltas, that of the Nile is mobile, shifting and sloshing about within its bedrocked chamber as ever more sodden northern African mud is piled on its surface, both above and below sea level. It is not a trivial matter. Every day, on average, some 330,000 tons of mud is dumped by the waters of the Nile somewhere in its delta.[15] Periodically, some of the steep-sided edges of the delta collapse under the huge weight of this, sending high waves backwashing across the delta lowlands. During floods, when water breaks out of the delta distributary channels, it spreads across these lowlands, leaving mud everywhere – unwelcome at the time but in the longer term essential for replenishing nutrients in the soil, ensuring it will continue producing abundant harvests to feed Egypt's growing population.

Thousands of years ago, the Nile Delta was well known throughout the eastern Mediterranean, the richness of its food production and its location as a gateway to Africa attracting the attention of countless expansionist tyrants and entrepreneurial merchants. The lands of the pharaohs were buffeted by incursions by Greeks and Romans, keen to secure footholds in fruitful Egypt.

Now consider Troy, not an Egyptian city, but north-west of the Nile Delta across 1,000km of the Mediterranean Sea and location of one of the most famed wars to have taken place in the period of written history, perhaps 2,700 years ago. Troy was the centre of a Hittite state that experienced almost constant friction with the Greek states on the other side of the Aegean Sea over the control of the trade routes therein. Friction turned into all-out war when Paris, scion of Troy, abducted Helen, Queen of the Greek state of Sparta, spouse of Menelaus and – by almost every contemporary account – the most beautiful woman in the world. Led by Agamemnon, the Greeks laid siege to Troy for a memorable 10 years until its inhabitants finally succumbed. 'Blood ran in torrents, drenched was all the earth,' noted Quintus Smyrnaeus.[16]

A detail often overlooked in this dramatic story is that when Paris and Helen first fled Sparta for Troy, their ship was blown off course and they ended up in the Nile Delta, finding sanctuary in a temple dedicated to the Greek hero

Heracles (Hercules). But rumours of their presence there reached Proteus, the ruler of Memphis, the principal city of Lower Egypt, who had Paris and Helen brought to his palace at the head of the delta. Characteristically, Paris tried lying[17] but the truth finally came out, whereupon Proteus announced his decision.

'Were it not that I count it a matter of great moment not to slay any of those strangers who being driven from their course by winds have come to my land hitherto, I should have taken vengeance on thee ... For thou didst go in to the wife of thine own host; and even this was not enough for thee, but thou didst stir her up with desire and hast gone away with her like a thief ... Now therefore depart ... This woman indeed and the wealth which thou hast I will not allow thee to carry away, but I shall keep them safe for the Hellene [Menelaus] who was thy host, until he come himself and desire to carry them off to his home; to thyself however and thy fellow-voyagers I proclaim that ye depart from your anchoring within three days and go from my land to some other; and if not, that ye will be dealt with as enemies'.[18]

Paris duly scuttled out of Egypt while Helen was held by Proteus to await her husband's arrival. But when Menelaus did arrive, his ingratitude so inflamed Proteus that he in turn caused Helen to find her way to Troy where she was reunited with Paris, thereby sowing the ground for the lengthy siege. Ten years later when Troy was finally overrun by the Greeks, Paris was long dead, Menelaus was about to kill Helen, but relented at the last moment – it is said on account of her persistent beauty – and took her back to Sparta.

The purpose of relating this story is because the temple of Heracles to which Paris and Helen first fled – and which

Herodotus, the author of the quotation, visited around 450 BC, some 250 years after the siege of Troy – no longer exists. The temple was part of the city of Herakleion, which disappeared about a thousand years after the time of Herodotus and was located only a decade or so ago, allowing its triumphant return from the realm of myth to that of reality.[19]

It goes without saying that the Nile river valley, with its myriad astounding human creations from times long past, has for hundreds of years been a drawcard for treasure hunters and archaeologists, their natural successors. Yet one aspect of the Nile's ancient human past that was largely overlooked until recently were the stories about large cities like Herakleion said to have once stood near the mouth of the river.

The Nile Delta constantly changes its form; river distributaries migrate from side to side and sometimes even jump from one place to another, slicing new channels through the unresisting alluvium and leaving old ones dry. Old channels may remain recognisable for decades, but any riverbank settlements that owed their prosperity to the proximity of the river must relocate elsewhere or decline. History shows that delta peoples are more accustomed to moving than people living on firmer ground, which is why deltas are such rich places for making discoveries about human pasts. If you can stand the stickiness and the stench of ancient muds, then you might reap rich rewards.

Today, the most important river channels in the Nile Delta are the Damietta and the Rosetta, but in the sixth century BC, more than 2,500 years ago, these were of minor importance, if indeed they existed at all. At that time the largest distributary channel was the Canopic, located in the west of the modern delta (Figure 3.3). Trade between the Egyptians and other peoples in the eastern Mediterranean was mediated by two delta cities – Herakleion and Eastern

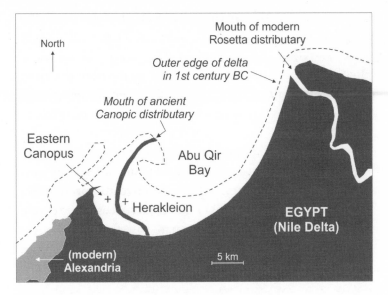

Figure 3.3 *Ancient submerged cities of the Nile Delta.*

Canopus – built along the banks of the Canopic. Herodotus tells us that 'if a stranger entered any other mouth of the Nile, he was compelled to take oath that he had come thither against his will; after which he had to sail his ship back to the Canopic mouth'.[20] Taxes were to be paid at these two cities for any goods being either offloaded or carried further upstream.

Herakleion was the leading commercial port for much of what we now call Egypt until it was displaced from this role in 331 BC by the founding of Alexandria. When the temple of Heracles was newly built, Herakleion also became a place of pilgrimage for people from throughout the eastern Mediterranean. But as time went on and as sediment built up on the floor of the Canopic channel, hindering navigation, so both Herakleion and Eastern Canopus began to acquire a reputation of a different kind.

In his *Geographica*, the Greek writer Strabo, who visited Eastern Canopus in 26 BC – considerably more than 2,000 years ago – noted that it too contained a famed temple 'which is honoured with great reverence and effects such cures that even the most reputable people believe in it'. Then, one can almost feel Strabo draw a long breath through his aquiline nose and purse his lips in disapproval before continuing.

> 'But to balance all this is the crowd of revellers who go down from Alexandria by the canal to the public festivals; for every day and every night is crowded with people on the boats who play the flute and dance without restraint and with extreme licentiousness, both men and women, and also with the people of Canobus [Eastern Canopus] itself, who have resorts situated close to the canal and [are] adapted to relaxation and merry-making of this kind.'[21]

Subsequently both cities disappeared. Herakleion in the east went first, vanishing in the first century AD, while Eastern Canopus in the west lasted until the eighth century. For hundreds of years thereafter, the stories about them by writers like Herodotus and Strabo were viewed as a bit suspect, perhaps based on erroneous geography, possibly invented – for how else might such great cities have been erased so completely from the face of the Earth? But along with the reputations of their chroniclers, both cities were restored to life in 1999 when archaeologists discovered their remains beneath the murky waters of Abu Qir Bay.[22]

Since then, much more has become known about these cities, their functions and their histories. Of key interest here are the reasons for their submergence, for it is not feasible to suppose such large cities were simply washed away in a massive flood or a tsunami, as once proposed, because their remains are largely intact. Many authorities, no doubt inspired by accounts of earthquake-linked destruction of cities elsewhere, have speculated that Herakleion and Eastern

Canopus met their fate in this way, but there is likewise no evidence for this, and the truth appears more mundane. For it seems that the cities became submerged as a combined result of both slow gradual subsidence and occasional bursts of rapid subsidence.

Slow subsidence typically affects delta cities the world over. It is one reason why cities like New Orleans (USA) and Shanghai (China) are experiencing problems at present. Slow subsidence continues to affect Nile Delta cities like Alexandria and Port Said, and is almost certainly due both to sediment compaction and to the downbowing of the Earth's crust under the weight of the sediments that comprise this mighty delta.[23] Yet implicit in the disappearance of Herakleion and Eastern Canopus is also more rapid collapse. For within their underwater sites today are the telltale signs of liquefaction, as would have been caused when the weight of the underlying sediment pile became just too great to continue standing upright and collapsed abruptly.

A few metres of subsidence was all that was needed for these cities to be abandoned – and eventually for memories of their existence to vanish from the minds of the delta peoples. A story your grandparents once told. Obviously untrue.

Not all such lost cities are found in deltas. Some sank or were washed away for other reasons. Close to the southernmost tip of the Peloponnese in Greece, just where the island of Elaphonisos lies offshore, are the remains of Pavlopetri, an astonishingly well-preserved underwater city. When it was initially surveyed beneath 4–5m of ocean, Pavlopetri was considered Mycenaean in age – a contemporary of Egyptian Herakleion – but with the discovery of Bronze Age and even a handful of Final

Neolithic pottery fragments, it is now clear that the site is likely to have been established much earlier, about 5,000 years ago.[24] This makes it more or less what the tourist brochures claim – the world's oldest submerged city.

Today we would not term ancient Pavlopetri a city, but during its heyday it was clearly at the upper end of complex populated places based on its extent, likely number of inhabitants and multiple functions. Most buildings within ancient Pavlopetri were stone-built and their forms can be traced today. Many had courtyards and two storeys, and were laid out along the side of a 40m-long street that linked with intersecting streets. Part of Pavlopetri appears to have been washed away by the sea so we estimate its maximum extent may have been some 80,000m².

Pavlopetri was one of the earliest port cities in Europe to thrive from cross-ocean trade and to exploit its associated economic power. Its ships helped move goods around the Mediterranean, specialising in exports of olive oil, wine and wheat in large locally made terracotta jars called *pithoi*. Although interacting with the Minoan peoples based on the island of Crete (and elsewhere) to the south, Pavlopetri never became a part of the Minoan Empire. Rather we can picture its people keeping up a brisk trade with Crete, importing a range of distinctive Minoan vessels. Intriguingly, it seems that the potters of Pavlopetri made cheap imitations of high-value bronze jugs from Crete, possibly to trade them for inflated prices, a distant echo of the modern trade in faux designer goods.

The end of Pavlopetri was not sudden. Steadily rising sea level in the eastern Mediterranean gradually eroded its viability as a port city as its lower-lying parts became infilled with sediments and flooded with increased regularity. While this situation is typical of many Mediterranean coastal cities from the times they were established right up to the present day, the site of Pavlopetri was also vulnerable to abrupt vertical movements of land associated with large earthquakes of the kind to which much of the eastern Mediterranean is periodically subjected. Frustratingly, we do not yet know which earthquake (or earthquakes) may have caused the sudden submersion of Pavlopetri, but there are abundant nearby analogues.[25] Yet it seems clear that, within a few hundred years of its prime around 3,000 years ago, Pavlopetri was gradually abandoned. By 2,000 years ago, the remains of the city were under 2–3m of ocean, although the isthmus between the mainland and offshore Elaphonisos remained passable by cart; some 2,000-year old cart tracks are shown in the image in the colour picture section.

Almost 9,000 years ago in the corner of a stone-walled house on the easternmost shore of the Mediterranean Sea, a woman coughed. It was different from the coughs to which she was accustomed but, having no choice, she carried on with her life. The next day for the first time ever, she coughed up blood. She told her husband, recently returned from trading cattle with inland peoples, and he told her that he had seen this sickness elsewhere. The woman was part of the community of Atlit Yam, within which it has recently been demonstrated that tuberculosis was present – the earliest-known incidence of this disease in humans. Once thought to have been passed to humans from cattle, which can also contract tuberculosis, new research shows unequivocally that the human pathogen is much older.[26]

Lying today a kilometre or so off the coast of Haifa beneath 8–12m of ocean, Atlit Yam was a coastal community that sustained itself mainly by fishing.[27] But it was more than this and research has given us extraordinary insights into the livelihoods followed by people in this part of the world 9,000 years ago. For a start, diets were more diverse than you might imagine. The people at Atlit Yam were not only fisherfolk, but also herders and hunters, cultivators and gatherers. They ate wheat and barley, lentils and flax seeds. They had domesticated animals including cattle and goats and, inevitably, dogs but they also hunted deer and pigs.[28]

Dietary diversity would have optimised food security, allowing the people of Atlit Yam time to construct houses, roads, wells and even rubbish tips across a core area of 40,000m^2, as well as perhaps freeing up time for other activities. These are hinted at by a number of likely ceremonial features at Atlit Yam including a circle of stones possibly used in a water ritual.

Atlit Yam was established around 9,300 years ago and, unlike Pavlopetri, met a sudden and unexpected end about 8,340 years ago. However, not all was smooth sailing in between. Like many similar coastal settlements across the world, Atlit Yam was founded in a time when post-glacial sea level was rising, and there is evidence that in the period it existed parts of it relocated inland and upslope on several occasions.

That fateful day 8,340 years ago began like any other. The people of Atlit Yam were going about their usual business. A hunting party was preparing to leave and the dogs gathered expectantly at the gates of the village. In another part, grain was piled up awaiting processing. A large stack of recently gutted fish stood alongside, ready to be cooked. But that never happened.

Without warning, a huge wave arose from the sea, 100m distant, and washed over Atlit Yam, followed within the next 10 minutes by a series of equally massive waves that brought an immediate end to all life in the village. This series of waves – a tsunami train – not only repeatedly flooded Atlit Yam, but also deposited layers of mud across its surface, fortuitously preserving some of its remains. Within and beneath these layers of stiff mud can today be found the bones of the people (at least 26) and the animals drowned in this catastrophe, as well as dispersed piles of gutted fish remains and grains. A key site within the ancient village is the former water well that became filled with clays mixed with a chaotic assemblage of human, animal and fish bones, explainable only as a result of great waves sweeping through the village and dropping a selection of the materials they carried into deep hollows like the water well.

That is not quite the end of the story of Atlit Yam for, incredibly, it has been possible to identify the origin of the tsunamis. They came from a collapse of the flank of Mount Etna in Italy, independently dated, which caused a massive displacement of ocean water that crossed 2,000km of the Mediterranean Sea within hours to reach Atlit Yam. Some of the waves that washed across it may have been as high as 10-storey buildings.[29]

While sudden rises of the ocean surface explain the end of Atlit Yam and may be implicated in the submergence of Pavlopetri and Herakleion, slower and more sustained rises in the ocean surface have caused the drowning of most submerged coastal places. In many instances, we collectively remember enough about what happened to know why it did and how events unfolded. Aside from John Smith's journal, there are eyewitnesses alive who can testify to the loss of land along the shores of Chesapeake Bay. Elsewhere, memories of land loss have been encoded within culture, be these the Pacific Island stories about people living on lands beneath the sea or the English tales of tolling bells from drowned church steeples.

But human memory alone can be a fickle fleeting thing. Recollections of even the most dramatic events in the past are liable to fade as these become ever more distant, particularly in places where there has been repeated turnover of culture. This involves the arrival of new people, strangers from elsewhere, who eventually amalgamate with the existing inhabitants and sometimes subsume their stories, even their associations with a place, into their own traditions. This is where science is sometimes able to help. We may still be unsure precisely where ancient Vineta was located, but the discovery by archaeologists of the types of structures and artefacts that might have been part of this medieval city strengthens the case for its existence. And consider the submerged cities of the Nile Delta, Herakleion and Eastern Canopus, forgotten until their presence was uncovered from beneath the sea in Abu Qir Bay. There is no question that Pavlopetri existed, but had it not been for its scientific investigation we would not understand its significance in Neolithic and Bronze Age Greece. And of Atlit Yam, we would likely know nothing today were it not for fortune and the doggedness of marine archaeologists.

This is not to say that science has all the answers. It is fair to say that Western science – which dominates understanding in the world today – is innately conservative and often exclusive (non-inclusive): traits that explain its success in many areas but also its stiffness and rigidity. Some modern scientists might be uncomfortable defending the words of Charles Darwin, who was key in developing Western scientific methods, when he said: 'I am a firm believer, that without speculation there is no good and original observation.'[30] It is that word 'speculation' that unsettles many scientists, something that seems to be anathema to them and the scientific method. Yet to me, it is speculation that marks the difference between science that is pedestrian and lacking innovation, and science that could prove a game changer.

CHAPTER FOUR

At the Nexus of Science and Memory

There are underwater places about which stories are fragmentary, suggesting those places may be imagined rather than real. Yet the incompleteness of the narratives could also be because these places were submerged so long ago that the associated memories have been broken up by their passage through time, rather like a balsa wood raft being swept through a rocky gorge. Bits emerge from the mouth of the gorge, but you cannot always be sure if they are flotsam or natural debris.

Such a place is 'that martyr'd land' in Cardigan Bay off the coast of Wales, where the city (or cities) of Cantre'r Gwaelod is reputed to have once stood.[1] A walled complex, likely to have been under threat from ocean waves for some time, Cantre'r Gwaelod (the Lowland Hundred) was the seat of King Gwyddno, from which he ruled Maes Gwyddno (the Plain of Gwyddno) covering an area of more than 2,000km². One night, the tide rising, the king's steward Seithennin, whose duties included checking all the flood gates were closed, became drunk and left one open. The ocean poured in, drowning the city and forcing its abandonment.[2]

Some scholars consider such stories to have been invented, which is of course possible although one might reasonably question why.[3] Perhaps we should instead take the essence of this story, shorn of its nativist and narrative fleece, and evaluate its credibility. People lived in what we now call Wales and Ireland 10,000 years ago and more,

so they would have witnessed the rise in the ocean surface
that eventually drowned the land bridge which connected
them during the last ice age.[4] This is undeniable so the only
credibility gap is whether or not we consider that extant
stories about the effects of that sea level rise on coastal cities
like Cantre'r Gwaelod might dimly recall these events or
not. You can stamp your feet and say 'absolutely not' (and
how presumptuous of this *eachtrannach* to try and teach us
the meaning of our traditions), but I think it is more prudent
to say 'perhaps' and dig into the evidence on that basis,
unfettered by partisanship.

As with many of the 'drowned cities' off the modern
coast of northwestern Europe, the exact location of Cantre'r
Gwaelod is unknown. This is often highlighted as a reason
for believing the story to be invented, but that of course is
a facile argument. If a place called Cantre'r Gwaelod existed
on the fringes of emergent Cardigan Bay when the ocean
surface was 20m lower than it is today, then consider that
not only have many millennia passed, many waves
undermined and dispersed its remnants, but also that tonnes
of mud washed off the Welsh hills, especially when they
were being rampantly deforested, have become plastered
over its surface obscuring its remains. Why should we
imagine it would be easy to find?[5]

Stories from the coasts of northwestern Europe that
recall submerged places are especially plausible. For not
only is it indisputable that human settlements have been
submerged along these coasts, but there are two other
factors that enhance the credibility of 'drowning' stories
here. The first is that, for reasons detailed in Chapter 8,
unlike almost every other stretch of coastline in the world,
many in northwestern Europe have experienced effectively
unbroken sea level rise since the end of the last ice age; in
most other parts of the world, this rise ended more than
5,000 years ago and sea level stabilised. So, stories about

'cities' being 'drowned' here within the past few millennia are more plausible than they are elsewhere. The second is that, compared with many other parts of the world, cultural continuity in Breton and Celtic worlds has been prolonged. In other places, cultures and their defining stories were regularly overturned, reseeded and replaced as new peoples arrived in particular areas, but few parts of the northwesternmost fringes of Europe have been so affected. Therefore the possibility that memories of drowned cities have been preserved here for thousands of years is also greater than elsewhere.

Back to Cantre'r Gwaelod. Imagine many thousands of years ago when Cardigan Bay was all dry land, productive and well managed by the yardsticks of the time, that the shoreline began moving inland.[6] The economy of the market centres (aggrandised to cities in some narratives) started to be affected, the rulers of the land becoming sufficiently concerned to build a series of sea defences to prevent inundation. At first, these were earthen dikes, later they became more resistant rock-faced structures. One of the first towns to be threatened – perhaps a place known as Cantre'r Gwaelod – sought to resist inundation at high tides and from storms by raising its surrounds, later by building walls and finally by introducing sluice gates to allow water that had overtopped (or underflowed) these walls at high tide to drain back into the sea at low tide. Maybe in a fit of mead-fuelled abstraction Seithennin did leave a gate open, but it was inevitable that the town would eventually be overwhelmed by the sea, whose rise was to prove unstoppable here. The people of Cantre'r Gwaelod and the surrounding countryside were forced inland, leaving Maes Gwyddno covered by ocean, the memory of when it was otherwise fading across the generations, kept alive only through the instructive story of Seithennin's profligacy.

Incidentally, the parallels with today are clear. As explained in Chapter 13, sea level is rising and likely to continue doing so for some time. We build sea defences, we blame any number of things for our situation, but eventually those of us living in low-lying coastal areas will have to move inland and upslope. In 5,000 years our descendants may struggle to reconstruct the geography of the early twenty-first-century world much as we struggle to reconstruct that of Cardigan Bay in the time of Cantre'r Gwaelod.

At the southern tip of India, the heartland of Tamil culture, poets and storytellers recall the former existence of a land known as Kumari Kandam. A vast, some say continent-sized, landmass reputed to have existed once in the Indian Ocean. Even though modern science has failed to locate it, the geography of ancient Kumari Kandam is remarkably well known from these ancient stories. Here is an example.

> O Indian Ocean! Where did you conceal our Tamil Nadu [Kumari Kandam], our ripe old land? Why did you plunder the forty-nine Tamil territories thousands of years ago? Where is that fine ancient river Pahruli? Where is that golden stream called Kumari? Where is the incomparable Mount Kumari filled with beauty? Where is [the city of] Then Madurai famed for the first Tamil academy over which presided the Lords Shiva and Murugan? Where is the lofty [city of] Kapatapuram, seat of the second academy graced by Tolkappiyar, Agastya, and many others? Where is Elutengu Nadu [Seven Coconut Territories]? Where is Elumunpalai [Seven Front Sandy Tracts]? Where is Elupinpalai [Seven Back Sandy Tracts]? And Elukunra Nadu [Seven Hilly Territories]? Where did these go?[7]

Tamil traditions about the 'lost continent' of Kumari Kandam have been used to support the science-derived proposition that a 'sunken' continent named Lemuria, discussed in the next chapter, exists in the Indian Ocean. It does not, but in an age before we realised that the Earth's crust was mobile, the distribution of terrestrial plants and animals (like lemurs along Indian Ocean coasts) seemed to require that now-submerged 'land bridges' and ocean-centred continents once existed. The imagining of such places was supported by stories like those about Kumari Kandam, with answers to questions about their existence, even their extent and the timing of their disappearance, reinforced by mutual cross-referencing.

Rather than having been a now-submerged ocean-bounded continent in the centre of the Indian Ocean – incidentally a geophysical impossibility[8] – it seems more likely that stories about Kumari Kandam recall the inundation of vast tracts of land now lying off the southernmost tip of India (Figure 4.1). This area is some $10,000km^2$ and was progressively drowned as sea level rose slowly across it in the wake of the last ice age, greatly affecting the people living there. Which is exactly what ancient Tamil texts describe.

Tamil culture is renowned for its poetry and literature and there is reputed to have been a succession of *sangam* (schools) established on Kumari Kandam.[9] The earliest of these was established in Then Madurai but, after this city was 'swallowed' by the sea, the *sangam* was re-established – presumably along with other capital functions – in Kapatapuram. This *sangam* is said to have endured for 3,700 years until Kapatapuram was destroyed by a 'deluge' after which it shifted to modern Madurai, where it remains (Figure 4.1). In details that appear unlikely subjects for invention, various Tamil texts recall how the seat of power represented by the Pandya monarchy also shifted in response to the progressive inundation of Kumari Kandam and

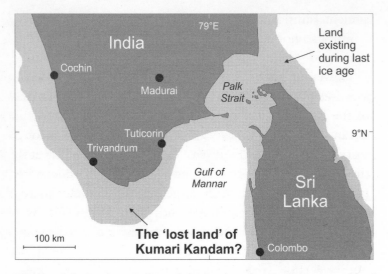

Figure 4.1 *Southern India and Sri Lanka, showing the extent of land loss in the past 20,000 years since the coldest time of the last ice age.*

describe where refugees from these 'drowned lands' were resettled.

Stories like those about Cantre'r Gwaelod and Kumari Kandam are likely to recall the effects of slow submergence of once much more extensive coastal fringes. Places that, owing to their fertility (for growing crops) and their flatness (for ease of access and construction), are likely to have been densely populated by the standards of ancient times. The gradual loss of these valued lands was bemoaned. A Welsh elegy about Cantre'r Gwaelod written by Guto'r Glyn around the year 1450 notes 'the lament of [King] Gwyddno … over whose land God turned the sea'. Tamil culture is likewise full of sadness, even recrimination, about the loss of Kumari Kandam and the associated degradation of Tamil culture by the 'cruel sea'. Addressing the ocean, a twentieth-century Tamil commentator noted that 'if today our ancient Tamil land and ancient Tamil learning had survived … if it had not been for your actions, we would have ruled the world'.[10]

How did such ancient submergence occur? Was it localised? Was it universal? In brief, during the coldest time of the last great ice age – around 20,000 years ago – the ocean surface was on average 120m lower than it is today, a situation shown in Figure 8.1. Most coastal dwellers at the time lived on lands that are now a long way underwater, evidence for their presence largely unknown. But there is no doubt people lived there and their descendants were forced inland by the rise in sea level that followed the end of the ice age, raising the ocean surface that same 120m within 10,000 years. If stories like those about Cantre'r Gwaelod and Kumari Kandam are based on observations of this sea level rise, then clearly they must have been passed down, largely by word of mouth, across hundreds of generations to reach us today.

That idea provides pause for many. A spoken story kept alive in intelligible form for several *thousand* years? Passed down across *hundreds* of generations by people who could not read or write? Instinctively, for a lot of people, this does not seem right. What such a view implies – and I have heard it expressed countless times – is a prejudice against orality, not something that can be denied on the basis of fact. Consider that during the last great ice age, when the ocean surface was an average of 120m lower than it is today in every part of the world, people were living almost everywhere except the Americas and the polar lands. There is no question that people *observed* the Earth when the sea level was much lower than today and that their descendants – our ancestors – survived the subsequent rise of sea level that radically transformed every coastline in the world. The question that may trouble you, as it troubles many literate people, is whether non-literate people could really have passed down their memories of these times across millennia. Yet they did.[11]

As antiquarian Angus Smith, aboard the yacht *Nyanza*, approached the Scottish island of St Kilda in 1872, he was both awed and unsettled at the sight. 'Had it been a land of demons, it could not have appeared more dreadful, and had we not heard of it before, we should have said that, if inhabited, it must be by monsters.'

Sometimes referred to as 'the island on the edge of the world', St Kilda is an outlier within the Outer Hebrides group of islands. One of the larger, more central islands is Harris, 90km across the ocean from St Kilda today. Incredibly, there is a story, reported by numerous authors and linked to folk traditions on both Harris and St Kilda, that the two islands were once one. The earliest written account of this oral tradition is more than 300 years old.

After a visit to St Kilda in the year 1697, one Martin Martin penned a description of this remote island and its singular traditions, including an allegedly ancient story about a female warrior who once lived above West Bay (the foundations of her house may still be visible). Martin explained that 'this Amazon ... was much addicted to Hunting, and that in her Days all the Space betwixt this Island and that of Harries [Harris], was one continued Tract of Dry Land', an expanse across which she frequently loosed her greyhounds when hunting deer.[12]

What this story might mean is of course open to debate, but it does seem an unlikely subject for invention – and it need not be so. For when after the last great ice age, possibly as recently as some 8,000 years ago,[13] large areas of what is now sea floor between St Kilda and Harris were dry land and there were certainly people living throughout the area.

It is plausible to suppose that here, as elsewhere in the world, the effects of sea level rise over the ensuing millennia were so impactful, so traumatising for the area's inhabitants, that stories about this event became etched into folk memories – a few of which, like that of the St Kilda Amazon, have found their way to us today.

As we saw in Chapter 1, there is an abundance of such stories in Aboriginal Australian cultures, recalling times more than 7,000 years ago when the Australian coastline was further seawards and when the people of the land walked, hunted and built associations around places that are today sometimes tens of metres underwater. Take the example of the world-famous Great Barrier Reef, the trace of which tens of kilometres off the north-east coast of Australia was once the shoreline.[14] The Djabuganydji and Gungganyji peoples of coastal Queensland recall the time when their ancestors roamed these lands until the day a man named Gunya broke a taboo that caused the ocean to start rising across them. This drove people of many different tribes inland, forcing them to mingle.[15] Then Gunya led a small group up a mountain where they built a large fire in which they heated rocks. Rolling the heated rocks downslope into the rising waters is said to have 'succeeded in checking the flood'.

Unlike most ancient flood stories elsewhere in the world, all known stories about coastal drowning from Australia recall the rise of the ocean eventually halting – but not reversing. These stories are thought to be the clearest and most extensive body of extant eyewitness accounts of the time – more than 7,000 years ago – when sea level was rising in almost every part of the world as a result of the melting of land ice formed during the last great ice age.

Perhaps the finest Australian example of a now-submerged connection between landmasses is that of Bass Strait, which separates the Australian mainland from the island of

Tasmania. Bass Strait is today more than 200km wide, no different than in 1788 when the British unilaterally declared Australia its newest colony. To help justify this declaration and the government-sanctioned land seizures that followed, Australia had been declared by the British to be *terra nullius* – a land devoid of people – on the tenuous grounds that its Aboriginal inhabitants were nomadic, not holding legal title to any particular place. This of course is hardly surprising – a bit like an Aboriginal Australian today walking into the National Parliament in Canberra and pronouncing their ownership of it on the grounds that they (and no one else) hold customary ownership of it. What worth any number of legal documents to the contrary in such a situation?

The point here is that *terra nullius* was a doctrine that in 1788 and thereafter was painfully at odds with the real situation. There were already people in Australia before the British and clearly many had been there some time – about 75,000 years in fact. Nowhere illustrates this better than Tasmania, which had an Aboriginal population of perhaps 5,000 when Abel Janszoon Tasman first stumbled upon it on 24 November 1642, naming it Anthony van Diemen's Land. On 3 December, ashore for the first time, crew members reported hearing human voices but saw no one.[16] Smoke from fires made it clear to Tasman that the island was occupied, but that did not give him pause when the carpenter of the ship *Heemskerck* swam ashore to plant the Dutch flag and unilaterally lay claim to the island.

The presence of a sizeable number of people on an island (Tasmania) a couple of hundred kilometres off the much larger landmass (mainland Australia) from which they must have once come, posed a bit of a dilemma for the first Europeans to wonder about it, especially as Aboriginal Tasmanians at the time of contact appeared to have negligible maritime abilities. Later it was discovered that the dingo – the wild dog of Australia now known to have arrived on the continent considerably later than its first people – was everywhere in the country in 1788, except Tasmania.[17] Together these two puzzles made it clear that the first Tasmanians probably reached their island on foot at a time when the Bass Strait was dry land (as it was throughout the last ice age), a time predating that when the first dingo made a claw mark in the closest parts of Southern Australia.

The 'land bridge' that existed between Southern Australia and Tasmania during the last great ice age, when the ocean surface was some 120m lower, was no narrow sinuous feature as the name might suggest, but a broad expanse of windswept land (see inset map in Figure 4.2). People 'crossing' it would have been unaware they were crossing anything. It was only when the sea level started to rise, drowning large areas of the land bridge, that it became more bridge-like (see main map in Figure 4.2).

In the case of the Bass Strait, the final land connection between mainland Australia and Tasmania – represented by the dashed line in Figure 4.2 – would have been when the ocean surface was about 56m lower than today, a condition calculated to have last been attained some time between 11,960 and 12,890 years ago. No wonder the dingo never made it to Tasmania. By the time it reached this part of the Australian mainland around 3,300 years ago, the land bridge would have been long gone, the ocean crossing probably too long and too dangerous for any human to attempt.

There are fragments of stories that may recall when Indigenous Australians walked across the Bass Strait land bridge or at least when the sea started drowning it. One collected in 1831 from Tasmanian Aboriginal people tells that 'this island was settled by emigrants from a far country, that they came here on land, that the sea was subsequently

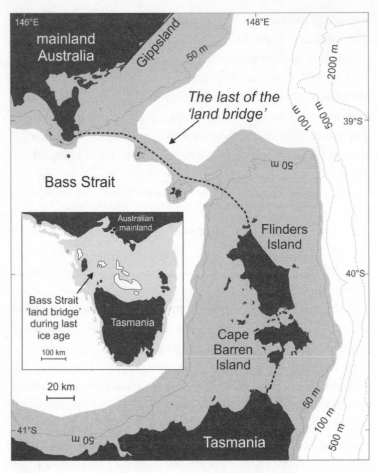

Figure 4.2 *The land bridge that connected Tasmania and the Australian mainland during the last ice age was submerged about 12,000 years ago, isolating Tasmanians for millennia.*

formed'. Another story collected from the Kurnai people of Gippsland recalls that:

> … long ago there was land to the south of Gippsland where there is now sea, and that at that time some children of the Kurnai, who inhabited the land, in playing about found a *turndun* [a musical instrument], which

they took home to the camp and showed to the women. Immediately, it is said, the earth crumbled away, and it was all water, and the Kurnai were drowned.

Unbeknown to the children, the *turndun* was to be viewed only by men so, when women saw it, calamity would necessarily follow. If these stories are based on memories from the time when the Bass Strait was still walkable, then the stories have to have been kept alive, largely orally, for almost 12,000 years.[18]

Ancient stories about submerged cities abound, especially in north-west Europe, where they include those of Ys in Brittany, France (discussed in Chapter 2); Lyonesse in Cornwall, England (discussed in Chapter 8); and as discussed earlier – Cantre'r Gwaelod in Wales. Stories describe the nature of the former 'city', the behaviour of the people occupying it and the reasons for its drowning. The memorability of such stories was periodically rejuvenated by the transference of critical details to new groups of culture heroes, key to why they have endured for millennia. And as discussed in the previous chapter, such stories have also been kept alive through cultural beliefs and protocols, ranging from the augural tolling of underwater bells to the requirements to bend your bare head to the spirits of drowned ancestors.

Confronted by these stories, the first reaction of most outsiders might be to dismiss the beliefs and the protocols, but we can look beneath these, treating them as expressions of cultural memories of distant events. Stories about lands that disappeared long before your great-grandparents' time might start to get a bit stale in the telling unless they are enlivened by improbable details – like sounds or scenes of human life beneath the ocean surface – or by complicated culturally grounded protocols that strengthen people's ties to particular places. A sense of loss is part of what defines us as sentient beings, able to feel and express emotion. I argue that many stories about submerged lands – in whatever cultural clothing they have become attired – represent a long-held sadness at the loss of place, something that can authenticate the actual loss of place. It is no longer good enough to tartly dismiss all such stories as inventions, as many scientists have done. Rather it is time to admit them to the cloistered realms of science, to look past their engaging surfaces and plumb their depths for evidence that these stories likely recall events that happened far earlier than we may instinctively regard possible. Such stories allow us insights not just into the facts of these events – in this case, the submergence of once-inhabited lands – but also into people's reactions to them.

All of this makes it relevant to consider the deep cultural roots of humanity's beliefs in sunken lands that were occupied by people who behave, speak and even look like us. Some of the latest manifestations of such beliefs include animated movies like *Atlantis* and *The Little Mermaid*, both of which are based on ancient stories. Plato's Atlantis never existed. As discussed in Chapter 2, it was created as a memorable vehicle for his ideas about the perfect society and how it might become undermined, but he based many details of his island-continent's catastrophic disappearance on accounts of real events that had occurred in and around the eastern Mediterranean at this time.

The existence of merfolk – and the lands under the sea we may imagine they inhabit – represents part of the modern residue of an older, more complex, altogether more sinister, set of largely European beliefs about what lay beneath the ocean surface. *The Little Mermaid* was a story written by Hans Christian Andersen who likely based it on Danish folk stories about underwater worlds. For example, one old story tells that a merman once appeared to a young woman living in Aarhus and enticed her to live with him at the bottom of the sea where they had many children. One day, hearing familiar church bells from the land above, the woman begged her husband to be allowed to visit Aarhus once more. He agreed as long as she returned, but she never did and it is said that his wails of anguish are often heard rising from the ocean depths nearby.[19]

Science has illuminated many of what were once considered profound mysteries about the deep ocean, so that crossing it by boat today no longer makes us as apprehensive as our ancestors must have felt making the same journey hundreds of years ago. In those days, fear of the ocean depths was partly assuaged by supposing them to be parallel worlds to the familiar ones above the ocean surface. This feeling was so strong in some cultures that people who were drowning would be left to do so, unaided, in the belief they were being pulled into the ocean by its subterranean inhabitants and that to resist this would be to incur their fury. Such sentiments might also be expressed by a person rescued from drowning, angered at their destiny being altered, slaying their rescuer.

In his 1822 novel *The Pirate*, loosely based on the activities of John Gow in Shetland, Scotland, Sir Walter Scott wrote of this practice. When Mordaunt, standing on the cliffs, saw a ship wrecked at sea and 'a man floating on a plank', he cried 'he lives, and may yet be saved'. Mordaunt rushed down the cliff path and tried to drag the man from the waves, but found he could not do so alone. He called out for help to Bryce, a local pedlar who was combing the beach looking for valuables from the wreck, but was ignored. Eventually giving up, he ran to remonstrate with Bryce who in turn shouted at him 'Are you mad … to risk the saving of a drowning man? Wot ye not, if you bring him to life again, he will be sure to do you some capital injury?'[20]

These beliefs are at such odds with modern ideas that it is tempting to dismiss them – from the mermaids to the drownings – as superstitions, built on ignorance, for which there is no excuse for believing. But we select the palatable elements of such ancient beliefs to entertain, surreptitiously brushing the rest into the garbage bin of history. This may be unhelpful because such beliefs, however ludicrous or distasteful they appear to us today, were once held by our ancestors who deduced them from their understanding of the world. Not for one moment am I suggesting that merfolk exist or that they inhabit undersea worlds uncannily similar to our own, but history does show – in countless instances – that such worlds have become submerged, the pain associated with their disappearance often rationalised by supposing they remain inhabited. The analogy with death, with the dead occupying a parallel realm in the 'heavens', has its roots in the same emotions.

I would argue that land loss has been a recurrent theme in the development of the human psyche and that many modern beliefs we hold, even ones we regard as fictional, have their roots in this. If you think this assertion bold, then consider the growing anxiety that many people today feel living on a warming planet where in many places, a century from now, coastal geographies are likely to have been profoundly altered, forcing us all to re-examine how to live in a changing world.

Whether they be poems describing the glories of sunken cities or bedtime tales about mermaids, many stories with ancient roots can be regarded as culture-defining, emblematic, often considered to be owned by particular groups of people and sometimes fed incompletely to outsiders as a consequence. Some stories of this kind may be labelled fictions, perhaps once a deliberate ploy to protect their essence from becoming known to others. The Neapolitan story of Niccolò Pesce, well known to every child growing up in this part of southern Italy, may be one such story.

Most versions of the story recall that young Niccolò was such an accomplished swimmer when he was growing up that it was clear his destiny would be one closely linked to the ocean. When he came to the attention of the King of Naples, he was charged with diving to the bottom of the sea, deeper than anyone had dived before, and reporting what he found. So happy was the King when Niccolò came back with stories of sunken gardens and treasures that he was dispatched time and again, even in some accounts returning with armfuls of jewels. Yet finally, unable to resist his monarch's blandishments to dive ever deeper, Niccolò vanished, presumed drowned but passing into the realm of legend. Some accounts have him reincarnated centuries later, occasionally appearing from the sea, shaggy and dripping, a knife in his right hand, to yarn with sailors in the Porto di Napoli (Figure 4.3).

The entire coastline around Naples has proved unusually changeable since it was first settled. Those parts within the Campi Flegrei (Phlegrean Fields) to the west of the city move up and down as liquid rock alternately fills and empties from comparatively shallow magma chambers below the Earth's surface. But most of the Chiaia Plain – along which the lower part of this great city is crowded – has a long history of slow sinking, largely attributable to nearby volcanic activity. Over the past few thousand years, successive generations of Neopolitans thus became inured to moving landwards, leaving the remains of their ancestors' dwellings under the sea, their associations passing gradually from memory into myth.

One of the most iconic sites along the coast of Naples is Castel dell'Ovo, built on an island named Megaris.[21]

Figure 4.3 *Bas-relief of Niccolò Pesce, Via Mezzocannone, Naples.*

Archaeological research on the ocean floor around Castel dell'Ovo has revealed much about its ancient past as well as the coastline changes that affected the ways people lived here. It also provides intriguing hints that stories about Niccolò Pesce may incorporate ancient memories about the nature of submerged places here.

The Greeks established the earliest settlement in the area about 2,700 years ago, but the first known use of Megaris Island was by the Romans, starting around 2,100 years ago, when a huge oceanside resort (*villa maritima*) was built there, one of a number (many now submerged) built at this time along the Chiaian coast as country retreats for Roman nobility. But since then Megaris has sunk and shrunk, its fringes forced back by the ingress of ocean water, much of its Roman heritage now anywhere between 2–12m underwater. It is known that many *villa maritimae* from this period involved major landscape modifications to allow the creation of ornamental gardens and orchards, connected by tunnels and *crepido* (paved walkways), as well as *piscinae* (fish ponds). Uniquely at Castel dell'Ovo in 6m of water, archaeologists found the remains of what they interpret as a *vivarium* (fish tank), dating from the first century BC and cut into the volcanic rocks along the island's western sides.

All these features are among those said to have been seen by Niccolò Pesce and it is quite possible that they were details, memories of ancient landscapes, that became attached to a story about a famed swimmer, living in medieval Naples, who drowned. If this is correct, then a later story can be seen as a vehicle, maybe intentionally selected, on which to fasten the disappearing details from eyewitness accounts of the fabulous Roman *villa maritimae*.

One final example of a place that sits squarely at the nexus of science and memory is that of Thompson Island, once apparently located about 72km north-northeast of Bouvet Island (Bouvetøya) in the central South Atlantic Ocean, a little-traversed region with few islands. No island exists today where Thompson Island is said to have been.

Why do we think that there was ever an island there, in this starkly landless part of the world ocean? The answer is because it was seen by people whom it seems reasonable to trust. The earliest sighting was on 13 December 1825 by the crew of the whaling ship *Sprightly* captained by George Norris, who named the island Thompson Island.[22] Its second (and last) sighting was in 1893 by Joseph J. Fuller and the crew of the schooner *Francis Allyn*. The absence of Thompson Island from its reported position was first noticed by the 1898 Deutsche Tiefsee-Expedition and later confirmed by the 1928–29 Norvegia Expedition, which criss-crossed the area.

Thompson Island was first seen at a time when the reckoning of longitude was less precise than it is today and some ocean scientists have simply dismissed its apparent sightings as having been Bouvet itself wrongly located.[23] Yet there is other evidence that Thompson Island may indeed have existed up until the last decade of the nineteenth century before being destroyed in a volcanic eruption, unrecorded because no one knew it had occurred.

A piece of evidence supporting this suggestion, admittedly one that 'falls short of conclusive proof', comes from contemporary meteorological records for Chile and New Zealand, which show an unexplained cooling between 1896 and 1907. Since this cooling episode does not show up in other records, especially from anywhere north of latitude 45°S, it is likely to have had a cause that was confined to the higher latitudes of the southern hemisphere. A sizeable volcanic eruption, of the kind that can cool temperatures for years because of the amount of fine particles it ejects into the atmosphere, would fit the bill.[24]

Sometime in the 1920s, the British Secretary of State for Foreign Affairs announced in the House of Commons that 'there did not appear to be any ground for questioning the existence of Thompson Island'.[25] That is a measure of how compelling the evidence was then regarded, but the lack of any subsequent scientific confirmation of its existence – perhaps the mapping of a volcanic shoal or an underwater mountain – is unfortunate. So, did Thompson Island ever really exist? Probably not, but just perhaps.

CHAPTER FIVE

Red Herrings: Fishy Tales of Unlikely Sunken Lands

Have lands, once inhabited by our ancestors, vanished beneath the ocean surface? There is no question about it. Yet for any sensible, open-minded person who wants to find out about such lands and the ways in which they vanished, there is hardly any material that is both accessible and credible. Not that there is any shortage of material as such, for this is a field that has become overwhelmed by a tide of nonsense. It is attention-grabbing, seeking to beguile readers with learned-sounding explanations of lost lands and their apparently key roles in the development of modern human culture that have eluded generations of closed-minded scientists.

Every part of the world has its myths, passed orally across countless generations, that recollect the disappearance of inhabited land, sometimes gradually, sometimes quickly. Yet human history is also full of falsehoods on this subject. Plato's story about Atlantis, discussed in Chapter 2, is an influential example. And yet the modern outpourings of new age and pseudoscience accounts of lost lands have met with conspicuous success, not least because they prey on the insecurities to which we are all subject.[1]

Non-scientific explanations of natural and physical phenomena are multiplying. To my mind, scientists have an ethical responsibility to try and counter these explanations or else risk that future generations will be ignorant and make judgements based on prejudice and misinformation rather than reason and fact.

In February 2006, I flew to Yonaguni, the westernmost island in Japan, a tiny outlier closer to neighbouring Taiwan than any of the larger Japanese islands. I was there to see the underwater Yonaguni Monument, claimed by a lone geophysicist and a gaggle of pseudoscientists to be part of a city occupied 10,000 years ago or more by an advanced race of people that conventional history somehow overlooked (it hasn't). In some interpretations, the city occupied the fringe of a now sunken continent – often called Mu – the breeding ground of the same advanced culture, spanning the entire Pacific Ocean (it didn't).

The sea was too rough for anyone to dive at the time of my visit so I hired a bicycle and cycled around the island, pausing at length to inspect the geology of the cliffs above the Yonaguni Monument. I am convinced the monument is entirely natural. The steps and the platforms that undeniably exist on it, making it look like a Mayan pyramid, are kept free of detritus and are smoothed by the movement of ocean water swirling around it, just as you might expect from a well-maintained artificial structure. But an examination of the cliffs along the south coast of Yonaguni, in which smooth rock steps and platforms also abound, convinced me that there is no evidence for human alteration of these and therefore unlikely to be any on a submerged structure made from the same rocks a few hundred metres away beneath the sea surface. The idea of the Yonaguni Monument having been fashioned by humans is a falsehood, as is the notion that it was once part of an ancient city, but it is certainly one that keeps the dive operators and bicycle hire shops on Yonaguni ticking over.

Another widely known modern pseudoscience falsehood about an undersea land concerns the Bermuda Triangle, an area in the western Atlantic where ships and planes have supposedly disappeared, repeatedly, in some interpretations being drawn down to a sunken city far below. The evidence is selective, distorted and easy to rebut, but this has not prevented any number of breathless books about the Bermuda Triangle once selling in their tens of thousands.

Another example concerns the *moai* (giant statues) on Easter Island in the south-east Pacific Ocean. Actually, these have been variously interpreted as representations of space aliens or the inhabitants of sunken Mu. So large are the *moai* and so backward were the people of the island supposed to have been at the time of their construction that explanations for the transportation of *moai* to the coast from inland quarries and their subsequent erection in lines along the coast have involved any number of asinine ideas. Aside from the usual crop of alien interventions, it has been supposed that the *moai* were raised by magic or some technology long since lost to the human race. Some New Age writers even suppose the *moai* wander the island today under their own power (they don't).

Such banal ideas might be considered harmless. But this is true only as long as they do not impede our understanding of the world, especially by people who are sincerely trying to make sense of it. If there is a heightened chance of aircraft crashing into the sea around Bermuda (there isn't), then science needs to have the data, interpret them and render appropriate advice. But the plethora of nonsensical explanations about such asserted phenomena often deter scientists from giving an opinion, even from taking an interest. This was commonplace a few decades ago when scientists were 'satisfied with offering incidental criticism or else ignore the problem [of pseudoscience] altogether', but now, perhaps emboldened by the conspicuous success of pseudoscience, more scientists are stepping up to counter its outrageous claims.[2]

As shown in the last two chapters, there are many stories about submerged lands that are verifiable or plausibly based, in part at least, on historical truth. But there are many other stories about such undersea lands that are false, although there is inevitably an excitable pseudoscientist pandering their duplicitous wares on every street corner ready to assure you otherwise. So too you can find a ubiquity of bearded panjandrums and bespectacled professors on cable TV willing to cast doubt on any evidence-based judgement you might be tempted to make about the authenticity of a particular tradition.[3] The whole business is confuddled by the way in which many ancient stories have become embellished so that their modern forms, in service of a particular agenda, may be grotesque caricatures of the original.

But we must make judgements about particular traditions. Those selected for discussion in this chapter are all considered to refer to submerged lands that never existed. Discussion of the most widely known of these – Atlantis – is omitted because, as explained in Chapter 2, it is clear it was the product of Plato's imagination.[4] Other traditions evolved in a variety of ways.

Let us start in the Indian Ocean. The ancient Greeks and the Romans knew the Indian Ocean existed, but their knowledge of its geography was far less than ours today. Their belief in a hemispheric symmetry of land and ocean – that landmasses in the northern hemisphere must be balanced by ones the same size in the south – gave birth to the proposition that a huge southern continent (later named *Terra Australis*) must exist. From the forays of Alexander the Great into South Asia, the southernmost piece of land known to the Greeks was Sri Lanka, which they regarded as an outpost of this great southern continent. This erroneous belief sowed the seeds of subsequent European notions about a lost continent in the Indian Ocean although, as discussed in the previous chapter, a Tamil tradition is more long-standing and probably based on authentic memories of land loss.

For Europeans, the next tranche of information about the geography of this part of the world came from the account by Marco Polo of his extraordinary journey to Cathay (China) in 1274 and his 17-year stay there. Part of his travelogue refers to a landmass he named Locac and described in inviting terms.

> When you leave Chamba and sail for seven hundred miles on a course between south and south-west, you arrive at two Islands, a greater and a less. The one is called SONDUR and the other CONDUR. As there is nothing about them worth mentioning, let us go on five hundred miles beyond Sondur, and then we find another country which is called LOCAC.
>
> It is a good country and a rich country; and it has a king of its own. The people are Idolaters and have a peculiar language, and pay tribute to nobody, for their country is so situated that no one can enter it to do them ill. Indeed, if it were possible to get at it, the Great Kaan [Kublai Khan, the Mongol Emperor] would soon bring them under subjection to him.
>
> In this country the brazil [nut] which we make use of grows in great plenty; and they also have gold in incredible quantity. They have elephants likewise, and much game … There is nothing else to mention except that this is a very wild region, visited by few people; nor does the king desire that any strangers should frequent the country, and so find out about his treasure and other resources.[5]

The problems with this account did not arise at the time, only much later, when nineteenth-century European geographers tried to identify the contemporary equivalents of the places Polo mentioned. All these geographers equated the place named Chamba with the island of Java, an error with monumental consequences for indeed, if you sail some 1,200 miles (1,930km) south-southwest of Java, you end up in the middle of the Indian Ocean. Therefore, it was deduced, there must have been a gold-rich continent named Locac in the middle of the Indian Ocean at the time of Polo's visit. It is not there today, therefore it must have sunk.

The reality is less exciting. Chamba is not Java. The distances and the sailing directions are all correct. The islands of Sondur and Condur are the modern islands of Pulo Condore off the coast of Vietnam, and Locac is modern-day Thailand. These errors were not rectified until the influential 1903 Yule-Cordier edition of *The Travels of Marco Polo* appeared, but by then it was too late. The genie had escaped from its jar, and any pseudoscientist happy to ignore anything published after the start of the twentieth century was likewise happy to embrace the story of the fabulous lost land of Locac!

The best known continent to have been imagined for one reason or another in the centre of the Indian Ocean is Lemuria. Inferred first on palaeontological grounds, this continent was so named by biologist Philip Sclater in 1864 who regarded it as an enormous land bridge, across which animals (such as lemurs) moved before it inexplicably sank. The existence of this once emergent, now sunken continent in the central Indian Ocean became the centrepiece of the famous 1868 treatise *The History of Creation* by biologist Ernst Haeckel in which he argued that Lemuria was the place where fossils demonstrating evolutionary links between apes and humans would one day be found. In a key section, Haeckel wrote:

The Indian Ocean formed a continent which extended from the Sunda Islands along the southern coast of Asia to the east coast of Africa. This large continent of former

times [is] called Lemuria, from the monkey-like animals
which inhabited it, and it is at the same time of great
importance from being the probable cradle of the human
race, which in all likelihood here first developed out of
anthropoid apes.[6]

Haeckel's role in the subsequent uses of Lemuria should
not be understated. For in an instant he transformed it,
with all the *éclat* of an internationally respected scholar,
from a place traversed only by various groups of non-human
animals to one in which humans had evolved to their
present condition and whence they subsequently spread
across the face of the Earth. Haeckel had no evidence for
this, but the influence of his view resonated through
subsequent ages and regularly resurfaces today in modern
New Age and pseudoscience writing.

   In retrospect, one might wonder whether Haeckel
intentionally rebadged himself as a maverick at this stage in
his life, allowing, as a few prominent scientists have done, an
urge for popular acclaim to override his scientific reputation.[7]
Many of his pronouncements about Lemuria lost him the
credibility his earlier career had earned him. For while
Haeckel originally postulated the existence of what came to
be known as Lemuria on the basis of quite plausible arguments
for the time, he got carried away with the possibility of a
huge sunken continent in the Indian Ocean by proposing
that it was 'the cradle of the human race'.

   The former inhabitants of this undersea land – the
Lemurians as they became – were, according to Haeckel, the
missing link between apes and humans. There is no evidence
for any of this but Haeckel's wild pronouncements, ironically
intended to support Darwin's ideas about evolution, became
fuel for some of Darwin's most vocal detractors and their
intellectual progeny. For example, the founder of Theosophy,
Helena Blavatsky, uprooted Haeckel's Lemuria from the

Indian Ocean for an allegedly sunken mid-Pacific location
where she considered humanity had evolved. We are the
fifth 'root race', according to Blavatsky, direct descendants of
the third root race, the first to occupy this Lemuria, and
who were four-armed egg-laying hermaphrodites, each
having an eye in the back of its head making them 'towering
giants of godly strength and beauty, and the depositaries [sic]
of all the mysteries of Heaven and Earth'.[8]

Now we move to the Pacific Ocean where there is no
shortage of stories about undersea lands. Some of these are
indigenous folk tales talking of such places as ancestral
homes, places of magic, abodes of dead souls. And then
there is a number of more recent stories best interpreted as
responses to the puzzlement felt by continental dwellers –
particularly from cold, crowded Europe – upon first
encountering an almost empty ocean covering nearly a
third of the planet's surface.

While there are innumerable formerly inhabited lands in
Pacific Islander mythology reputed to lie beneath the sea,
many stories refer to a single one – Hawaiki. For almost all
peoples residing in Polynesia (including Hawaii and New
Zealand), Hawaiki is a place of origin, a place from which
the ancestors came. In other parts of the Pacific Islands
region, there are lands that fill the same role. In parts of Fiji
and Tonga, for example, the land is Burotu or Pulotu; in
parts of the island nation of Kiribati, the land is Matang.

Hawaiki is mentioned so frequently in authentic Pacific
Islander myths that its key functions are comparatively easy
to determine. For many Pacific Islanders, Hawaiki was the
first land on which people ever lived. In New Zealand
Māori traditions, humans were metaphorically fashioned
from the soil of Hawaiki. It clearly adds to both the

believability and the sacredness of stories about Hawaiki for it to be no longer where it once was, no longer accessible, maybe even mysteriously vanished. For the enduring attraction of any such place – be it Heaven or Nirvana or Hawaiki – lies in its unattainability by mortals.

To a mythmaker, intangibility and unattainability are useful attributes for ancient places because they allow any number of fabulous functions to be unquestioningly assigned to them. Thus, in Pacific Islander traditions, Hawaiki is not only a homeland but also sometimes where dead souls reside. Many islands and island groups in the Pacific have traditions of the routes followed by souls after death. Invariably these routes end at the top of a cliff, off which dead souls jump in order to reach underwater Hawaiki (Figure 5.1).

Pacific Islander stories about Hawaiki laid the foundations for a merger between ancient Pacific Islander traditions and upstart Western pseudoscience. An influential passage was written in 1920 by John Macmillan Brown, one-time Chancellor of the University of New Zealand, who

Figure 5.1 *The route of dead souls across the Fiji island of Vanua Levu ends here at Naicobocobo Point, also known as icibaciba (the jumping-off place of the souls of the dead). At night people in nearby villages are sometimes awoken by dogs barking frantically at something invisible, said to be a dead soul on its way through the village to jump into the ocean at Naicobocobo Point [photo by author].*

described Hawaiki as 'a mythical and shadowy land beneath the waters'. This detail has been seized upon by any number of pseudoscience writers and New Age theorists to support the idea that, lurking within the deep dark waters of the Pacific, lies a lost land known to the region's earliest inhabitants as Hawaiki. This idea melds with the prejudices that many Europeans earlier developed about the Pacific, an ocean so vast and effectively landless – and therefore so alien to their experiences and beliefs – that its origin became a matter for speculation.

Put yourself in the leather boots of some of the pioneer European explorers of the Pacific. People like the irascible William Bligh, the reflective Jules Dumont d'Urville or workmanlike James Cook. Raised in continental worlds to uncritically regard these as normal, it must have been quite a jolt to traverse the vast Pacific Ocean and not encounter a single sizeable landmass. For such people, it must also have been fairly easy to impose the European norm of how the world should be on the anomalous Pacific by supposing that its isolated islands represented the highest parts of a continent, which looked like Europe, that sank.

Equally misleading was the common inference that Pacific Island peoples had not been in contact with each other since this supposed cataclysmic event. The manifest language similarities between people living on islands hundreds, if not thousands, of kilometres apart were interpreted as residual, even proof of the 'sunken continent' hypothesis, given that it was adjudged impossible that Pacific Islanders had the requisite seafaring skills to travel such great distances. But they did.[9] Before Europeans even knew the Pacific Ocean existed, Pacific Island peoples had crossed the entire ocean from west to east, a distance of at least 16,000km, settling most habitable islands they encountered and finally washing up on the western shores of the Americas in at least three places.[10]

There are also indigenous stories from the Pacific, unblemished by foreign revisionism, that are worth mentioning. For example, among the peoples of the atoll islands of the central Pacific, particularly those in Kiribati and in the Tuamotu Islands of French Polynesia, there is a wealth of stories about what might be termed ghost islands: in Kiribati, *abaia anti*. These are islands that alternately appear and disappear – sometimes inhabited, sometimes not – that wander around the ocean, never staying long in one place.[11]

Many *abaia anti* have no names. Of those that do, Te Bike (literally 'the sandbank') is most common. One island named Karawanimakin was located near modern Makin Atoll and vanished slowly enough to allow its few inhabitants to move elsewhere. When last seen, decades ago, it was a patch of sand, home to a few nesting birds. Now it is gone and no one is sure where it was – or indeed whether it really was. Further from Makin, one Te Bike was reportedly landed on by the crew of a trading vessel in 1908. Decades later, colonial administrator Arthur Grimble instigated a search for it, but never found it, although more recent information places it in the southern Marshall Islands close to Mili Atoll.

It is likely that the inherent fragility of certain atoll or reef islands in the face of strong winds and high waves led to such stories in these parts of the Pacific. There must have been many instances where a superficial island, a sand cay perhaps, known to people from nearby islands suddenly disappeared – while another, on a reef flat elsewhere, suddenly appeared. For many of the same processes that lead to the destruction of such frail islands sometimes also cause them to form. Waves smash into these islands, mobilising their constituents and dumping them all offshore, while elsewhere the same waves scoop sediments up from the sea floor and pile them on to shallow reefs to

create islands. And how would the people of nearby larger
and more stable islands, some occupied for more than 2,000
years, encrypt this important information in their oral
traditions? Speaking of ghost islands that move in and out
of the visible world is an explanation that is in keeping with
traditional Kiribati world views.

Abaia anti are elusive but – to most Kiribati people
(I-Kiribati) at one time – they were real enough to have
underwritten their world views. Kiribati traditional beliefs
hold that the visible material world is paralleled by a number
of similar worlds that are invisible and intangible. Yet these
worlds are not rigidly bounded and objects (like islands)
and people can move between them. When people die,
their spirits cross from the visible world into an invisible
one, but sometimes later return, not as the malevolent or
mischievous ghosts of western fiction, but as material
beings in their original forms whose presence is routinely
accepted by the islands' living inhabitants.[12]

It may seem irrational to equate the periodic appearance
and disappearance of rarely visited islands in particular places
with the idea that these islands actually wander around the
ocean, imbued with a life and purpose of their own. Yet it is
an idea that was current in many Pacific Island cultures a
few hundred years ago. In 2004, Tiata Koriri, an elderly man
from Matairawa on Makin Island, explained that once all the
islands in the northern part of Kiribati had wandered across
the ocean 'until one mighty ancestor held them down with
an anchor made from strong wood'. And in the Tuamotu
Islands of French Polynesia, islands were once said to wander
or float around, often fleeing as sailors approached them;
'the Vanishing-Isle is as a migratory bird flashing in
undeviating flight, now launched upon the wind', goes the
translation of a traditional Tuamotuan chant.

In Kiribati traditions, when islands vanish into an
invisible realm, they are often spoken of as being under the

sea, where indeed their foundations might really have gone. Yet the literal translation of the Kiribati term for this (*i nano*) is 'where the sun goes into the depths', which could mean either that the island is underwater (where sunlight penetrates) or that the island is over the western horizon (in the direction of the setting sun).[13] It is tempting to posit that the ambiguity was deliberate, intended to pull a cloak of mystery around these ancient tales.

Similar arguments can be made for the Hawaiian tradition of Kahiki.

O Kahiki, moku kai 'āloa,
Āina o 'Olopana i noho ai!
I loko ka moku; I waho ka lā;
O ke aloalo ka-lā, ka moku, ke hiki mai.

Kahiki, land of the far-reaching ocean,
Land in which Olopana dwelt!
Within is the land, outside is the sun;
When it arrives, the sun is favoured by the island.

These lines are from the *Chant of Kuali'i*, an epic poem well remembered in Hawaiian oral traditions around the start of sustained European contact in the mid-nineteenth century that is thought to have originated about 300 years earlier.[14] The poem describes an ocean journey made by a man named Kuali'i that took him to a place named Kahiki, the inhabitants of which were unlike those of Hawaii and spoke a language quite unfamiliar to him.

Given that the details about the voyage and the land Kuali'i reached are so considerable, it is possible that the 600-line poem is the record of a real voyage. Perhaps Kahiki was Tahiti in French Polynesia, the place from which the first Hawaiians are likely to have come hundreds of years before. Yet in Hawaiian, *kahiki* means 'the east' whereas Tahiti is to the south. This is where it becomes really interesting for no land lies east of Hawaii until mainland America is reached, so it is possible that Kahiki lay somewhere along the western seaboard of the Americas. This explanation would satisfy the detail that the people of Kahiki did not resemble those of Hawaii and spoke a completely different language, neither of which would have applied to Tahiti.

So how might this have happened? In one scenario, Kuali'i was fishing one day out of sight of land when a Spanish galleon, plying its ponderous way between the Philippines and Mexico laden with pepper and porcelain, spotted him. Both parties were equally astonished – the Spanish did not know of any islands in the area – but somehow Kuali'i climbed aboard the galleon and sailed with it to Acapulco. Here he spent some time before another trans-Pacific galleon took him back to familiar territory so he might tell the tale of his incredible adventure, one that became immortalised in the *Chant of Kuali'i.*

Of course, we may be reading too much into this. Kuali'i may simply have been the invention of a Hawaiian bard whose poem about Kahiki was nothing more than a tale about an imaginary land. At once a place of mystery and of hope, reassuring Hawaiians that other people lived beyond their horizons. That they were not alone on this isolated group of islands.

Now we turn to the Atlantic. At one time it was believed to be littered with islands, reflecting what historian John Gillis described as 'medieval islomania', most of which never existed.[15]

There are so many versions of the at least 1,500-year-old story of the discovery of the Island of St Brandan that it is difficult to know which was the original.[16] It seems clear that the story involves the discovery of many islands rather than just one and is rooted in ancient Irish sea sagas (*immrama*), particularly *The Voyage of Bran*. To understand the purpose of such voyages, appreciate that in western Ireland during the first millennium AD, the barely explored and seemingly boundless Atlantic Ocean was reputed to be the source of all kinds of devilry. What better test of the potency of the Christian message than for its bearers to enter the devil's realm and return to tell the tale?

The story goes that in the sixth century, Brandan, Abbot of Clonfert in Ireland, heard rumours about an unspoilt island somewhere off its west coast. So, in a wattle-and-hide boat, Brandan and 16 monks sailed south-west for 40 days, discovering a number of islands which, though mostly inhabited, had previously been unknown to them. Among the islands they visited was one that was the site of an erupting volcano and another that famously turned out to be a large fish. Obviously, Brandan and company lived to tell the tale.

Some analyses of this legend claim it to be true in outline, Brandan actually encountering islands in the Azores, the Faeroes or even off the east coast of North America that were unknown in his world at this time. But there is no proof of this and it is more parsimonious to regard the account of the voyages of St Brandan as an allegory of his success in spreading the word of God to faraway people. The possibility that the account of the voyage is that of a dream is certainly 'in keeping with the anchorite enthusiasm' of the time and should not be readily discounted.[17]

Another Atlantic island that probably never existed is Antillia. This is said to have become known to Europeans in the year 711 when, following a battle in which the invading Moors overwhelmed King Roderick of Spain's army, many Spanish and Portuguese refugees sought to flee the Iberian Peninsula. One group boarded three boats at Gibraltar and sailed in search of islands rumoured to exist in the 'Sea of Darkness' (the Atlantic). Guided by the archbishop Oppas, their target was the large island of Antillia, which he was charged to divide into seven parts and rule accordingly. Then …

> One morning, at sunrise, there lay before them a tropic island, soft and graceful, with green shrubs and cocoanut trees, and rising in the distance to mountains whose scooped tops and dark, furrowed sides spoke of extinct volcanoes – yet not so extinct but that a faint wreath of vapor still mounted from the utmost peak of the highest among them.[18]

The archbishop did as he was bid, resulting in the other name for Antillia as the 'Island of the Seven Cities'. But it has never knowingly been seen again. When Christopher Columbus reached the Caribbean island of Hispaniola on 5 December 1492, he gave the name of Antilles to the island group of which it is part, believing he might have found the legendary Antillia.

Perhaps the most intriguing aspect of the story of Antillia is the inference that Europeans knew of lands within the Atlantic, perhaps even along its western edge, long before Columbus sailed there from Spain in 1492. Indeed, contrary to long-held beliefs, it now seems that Columbus already had information, possibly even maps, of parts of the Americas before he set out for them. Whether that information came from eighth-century Iberian refugees or was a legacy of voyagers from China in 1421 (or both) is uncertain.[19]

The island of Hy-Brasil is not near Brazil, but was reputed to lie off the west coast of Ireland. It has been sometimes identified as Terceira Island in the Azores or maybe even an island adjoining the east coast of North America. But almost certainly it never existed in the way most versions of the story describe it. A common story is that the island of Hy-Brasil was discovered during the thirteenth century by the crew of a sloop returning from France on their way to Killybegs on the west coast of Ireland after they lost their way in thick fog, eventually beaching on an island overrun with black rabbits. They found the island to be extremely fertile and abounding in precious metals, samples of which they took home with them. Subsequent attempts to locate Hy-Brasil all proved futile.

But still the tradition survived for hundreds of years, fuelled first by an incomplete picture of the geography of the Atlantic Ocean and the mysteries it was perceived to hold, then later through sentimental attachment to an ancestral belief. For instance, in the early nineteenth century:

> The people of Aran [islands off the west coast of Ireland], with characteristic enthusiasm, fancy, that at certain periods, they see Hy-Brasail, elevated far to the west in their watery horizon.[20]

Having travelled around this part of the Atlantic, these Aran islanders had enough knowledge to know that Hy-Brasil did not exist, but rather than callously discard their forebears' beliefs they transform them. No longer do they believe that Hy-Brasil is a real place, but instead they consider it an enchanted island, hidden most of the time from mortal view, which only occasionally becomes visible. Its function is as a link to the narrative richness of the past.

Dogged readers of pseudoscience writer Graham Hancock who got past page 500 in his meandering 2002 tome *Underworld* will find he makes much of the possibility that Hy-Brasil is today marked by the site of Porcupine Bank, the surface of which now lies 55m beneath the sea surface off the west coast of Ireland. I was initially puzzled when I read all this because I could not relate it to the main thrust of Hancock's diaphanous arguments. Then it clicked. Hancock wanted to show that people had mapped Hy-Brasil as Porcupine Bank when this was emergent; in other words, people had been making maps of mid-ocean islands before Porcupine Bank became submerged about 10,000 years ago. Hancock wanted to do this because he was trying to demonstrate, contrary to all evidence and reasoning about the subject, that there was an advanced civilisation on Earth 10,000 years ago, remnants of which seeded the earliest modern civilisations. The fallacy in Hancock's reasoning is to suppose that any association between Hy-Brasil and Porcupine Bank would have to have been *mapped*. If Porcupine Bank was known to people 10,000 years ago, this knowledge would have been passed on *orally* – a tradition that could not have been mapped for at least another 8,000 years.

A final island – 'existence doubtful' (ED) as such places were once labelled on nautical charts – to mention from the Atlantic is Buss Island. In the mid-sixteenth century with all the dangers attendant upon reaching China from Europe by going around the Cape of Good Hope or Cape Horn, there was a flurry of activity to find a Northwest Passage from the Atlantic to Cathay (China) that would give European merchants a safe and fast route to this land of fabulous riches. Sir Martin Frobisher proved the most dogged of the searchers for the Northwest Passage, although he never found it. But on his third and last attempt in 1578, one of his 15 ships (a busse) had to turn back and came across

> … a great Ilande in the latitude of 57 degrees and a half, which was neuer founde before, and sayled three dayes along the coast, the land seeming to be fruiteful, full of woods, and a champion countrie.[21]

The island was named Buss Island and may have been sighted again in 1606 and 1671, but not subsequently. Belief in Buss Island persisted and, when it became certain that it was not where it had been claimed to be, it was thought to have sunk, an idea that received some support from depth soundings and observations of breaking waves in this bleak fog-drenched area.

Almost certainly Buss Island never existed. The claimed sightings were probably of a part of Greenland or of an island off the east coast of modern Canada that was mistakenly considered a new discovery. Certainly, we can reasonably disregard the 1671 report of Captain Thomas Shepard, an observer 'gravely distrusted' in one evaluation, who claimed to have landed on Buss Island, ingratiatingly naming 12 of its most prominent features after the directors of the Hudson Bay Company he represented.[22]

So why were stories about sunken lands invented? For a sensible answer, we might turn to Thomas Johnson Westropp, well-acquainted with the spoken stories he spent much of his life collecting in nineteenth-century Ireland. Westropp was not fooled by the sight of the island of Hy-Brasil off the coast of County Clare in 1872.

> The desire for the ageless, deathless land prevailed all up the western coast, and was strong in Kilkee in 1868–78, and perhaps even still. I myself saw the mirage several times in 1872 giving the perfect image of a shadowy island with wooded hills and tall towers springing into sight for a moment as the sun sank below the horizon. I have also heard from Kilkee fishermen legends … of men starting seaward to reach its fairy shores, and never returning.[23]

Life in rural Ireland was unbelievably hard during the
middle and late nineteenth century. Around a quarter of
the country's entire population perished during *an Gorta
Mór*, the great famine that lasted from 1845–52. In the
uncertain aftermath of such terrible events, it is not difficult
to understand why a belief in imagined lands of plenty,
which might have been considered risible at other times,
should become so strong.[24]

The Ireland situation is not without parallel. For in the
Pacific Islands, particularly the more remote ones where
food crises resulting from prolonged droughts periodically
occurred before the age of globalisation, we know
something about the responses of island societies. For
example, in former times in the Marquesas Islands (French
Polynesia), there were many traditions talking of the
existence of islands of abundance over the horizon. At
times of famine, canoes full of people set sail with the
declared intent of *he fenua imi* (land-seeking). One account
from 1930 spoke of more than 800 canoes that had at one
time or another set off on such futile voyages; of these 800
canoes, only one was ever heard from again.[25]

Today, when almost everyone in the world knows
better, sightings of imaginary lands are probably wholly
bound up with the long traditions about such fabulous
places and the often unspoken will of the community to
sustain them. This may be true of sightings of Kumari
Kandam off the coast of southern India that speak to the
heart of Tamil nationalism. It also seems a likely explanation
for the periodic sightings of Burotu Island off the coast of
Matuku Island in south-east Fiji, sightings sometimes said
to herald significant events, such as the imminent demise
of a chief or, more mundanely, a good omen for an
upcoming rugby tournament.[26] A verdant island also
sometimes appears off the eastern seaboard of the United
States near Boston, an event said to have once had people

scrambling for their boats to sail there, but never succeeding in doing so.[27]

Once we understand why people choose to believe in such lands, even to the extent of deliberately blurring the distinction between the real and the imagined, we then need to understand what phenomena may actually trigger such illusions. There is no shortage of optical phenomena, veritable tricks of the light, to explain the apparent sighting of such islands. Earthquake waves funnelled into passages through coral reefs may explain some of the sightings of Burotu. Other imagined islands, particularly those viewed from within the ocean rather than from some adjoining cliff top, may be distant cloud banks, overturned icebergs stained with algae, schools of breaching fish, thin yet coherent mats of pumice, dead whales or merely places where two sets of opposing waves converge.

Yet there are instances of islands where we simply do not know the truth. Was there once an island there that somebody saw, but which has since vanished? Or was the island only ever illusory, the fevered product of a sailor's imagination? Let me give you two examples.

In the southernmost Atlantic Ocean, between the Falkland Islands and South Georgia, no land is visible today. Yet in 1762, the whaling ship *Aurora* found a group of islands here, at 52°37'S and 47°43'W. Named the Auroras, these islands were sighted by five ships in the following 30 years. In 1794, surveyors on board the Spanish corvette *Atrevida* mapped and sketched three of the Auroras. Even though no one ever set foot on these islands, the case for their authenticity appeared 'ample and convincing'.[28] But then it all began to unravel. The Auroras could no longer be found. In 1820, James Weddell, the renowned Antarctic explorer, searched for them in vain, as did for 15 days in November 1822 the somewhat less-renowned

Benjamin Morrell (pictured in Figure 2.2). Today you will not find them on any chart.

Of course, one cannot help but wonder why none of the people who saw the Auroras ever landed on them, but is that a landlubber's suspicion? The whalers and sealers had a quota to catch, and little time to waste exploring barren icy islands.[29] To get that question better in perspective, consider the story of Los Jardines, an island that no longer exists but which was precisely located and reportedly landed upon more than 200 years ago.

Not remotely icy, Los Jardines is said to have been in the north-west Pacific Ocean, at 21°40'N and 151°35'E. The first and fullest report of the islands was made by the Spanish navigator Álvaro de Saavedra Cerón in 1529. Saavedra landed on the island, which he found to be inhabited. He thought

> … that the people had come originally from China, but had, by long residence, degenerated into lawless savages, using no labour or industry. They wear a species of white cloth, made of grass, and are quite ignorant of fire, which put them in great terror. Instead of bread they eat cocoas [coconuts], which they pull unripe, burying them for some days in the sand, and then laying them in the sun, which causes them to open. They eat fish also, which they catch from a kind of boat called parao, or proa, which they construct with tools made of shells, from pine wood that is drifted at certain times to their islands, from some unknown regions.[30]

Of the three subsequent sightings of Los Jardines, all in the same spot, that in 1788 by Captain William Marshall is most trusted for he was a competent navigator who accurately mapped many islands before this time. Los Jardines has not been seen since 1788. In 1940, the mapping of an underwater mountain peak, possibly a sunken island, close to their alleged position, convinced Captain G. S. Bryan of the United States Navy that Los Jardines had indeed once existed here but then sank.[31] Some scepticism comes from the fact that this peak is now more than 2km beneath the ocean surface, a decidedly yet not inconceivably long way for an island to drop within such a short time (see Chapter 10).

CHAPTER SIX

Hidden Depths: Coastal Lands Submerged Long Ago

Dogs are often difficult to train, but the lineage of the efforts to do so is extraordinary to contemplate, with recent research suggesting that people first domesticated dogs (from wolves) as much as 40,000 years ago.[1] Almost all dogs today have European genes, suggesting Europe was where canines were first domesticated, although demonstrating the when and the where is challenging.

So, let us go back a little less, a mere 8,000 years, and imagine the life of a dog in a part of Europe that is mostly underwater today. This dog weighed about 30kg, had the appearance of a husky and lived with a group of Kongemose people on one of the marshy islands dotting the sea around modern Denmark at the time. The dog was part of a pack, rangy and uncompromisingly competitive creatures, which slept beneath the Kongemose long houses, fighting over food scraps and bones, alert through the long dark winter nights when other creatures craving food might approach the village. Kongemose dogs ate mostly seafood, leftovers of the people's commonest meals. But occasionally the dogs would accompany Kongemose hunters to other, more inland places, where they would help catch roe and other deer, rewarded for their efforts with their less readily edible parts – dietary subtleties that can be deduced from isotope analyses of dog bones retrieved from the submerged remains of Kongemose settlements.[2]

Where the North Sea meets the Baltic Sea in north-west Europe, there are today thousands of islands, small and large, which represent the residue of the inundation of the region following the last ice age. Most of Denmark's 406 islands are today uninhabited, too small to sustain livelihoods – a comparable situation to neighbouring Sweden. Although the Baltic Sea was covered by a massive ice sheet 20,000 years ago, the coldest time of the last great ice age, much of this ice melted within the next 10,000 years and the region became covered by the vast freshwater Ancylus Lake. This proved a magnet for all manner of terrestrial species, including people, who first arrived here at least 9,000 years ago. But Ancylus Lake survived only another thousand years or so before the rising ocean poured across its western margin. Later, as the Earth's crust in the area rebounded upwards (a concept illustrated in Figure 8.2), there was another freshwater phase before the modern Baltic Sea came into existence a mere 4,000 years ago. You might think the people of this region, whose livelihoods for millennia involved hunting wild animals and gathering wild foods, might have become accustomed to changes in the environment. At some points that may have been true, but when there was rapid change, especially associated with rising sea level that 'drowned' coastal lands, trimming larger islands down to smaller ones, adaptation would have been harder.

The Kongemose Culture flourished 7,200–8,000 years ago in what today we call Denmark and southern Sweden, at a time when the ocean surface was rising rapidly, the last spurt of post-glacial sea level rise here. Within this 800-year period, sea level may have risen as much as 20m, an average of 25mm per year. Compare that with the global average rise of sea level at 3–4mm per year that we are currently experiencing and you can appreciate the magnitude of the environmental forces shaping the Kongemose world. An adult in their 60s might have witnessed the ocean surface rise 1.5m, a vertical shift that may have seen hundreds of metres shaved off the coastlines of their childhood. The region 'went from being a relatively contiguous land area to being divided up into a myriad of islands',[3] forcing changes to how Kongemose people lived. In earlier times, occupying larger islands, they used arrows and spears tipped with lethal mistletoe toxin to kill deer and wild boar in the forests of the region, but as the area of these shrank the Kongemose began to focus more on acquiring foods from the lengthening coastlines. They carved fish hooks from red deer bone and built cylindrical fish traps from willow twigs – they adapted.

The combination of the rapid rise of sea level during Kongemose time and people's growing preference for coastal foods – and coastal living – means most Kongemose settlements are now underwater. Figure 6.1 shows some of the numerous Kongemose settlements in what is today the Roskilde Fjord, a few tens of kilometres north-west of Copenhagen on the Danish island of Zealand. Some of these ancient underwater settlements have names (shown in italics); studies at Blak have been so thorough that they have allowed the uniqueness of the Kongemose Culture to become widely appreciated.

But reflect on what we might know about the Kongemose had we looked only on land for evidence of their existence. Hardly anything, perhaps nothing at all. We might even assume that this region was unoccupied at the time and that the earliest people to live there were the well-studied Ertebølle people, the successors to the Kongemose, after the ocean surface here ceased rising 7,200 years ago. We might also never have understood that the Kongemose utilised the services of domesticated dogs in this region. Dogs became the loyal companions of the Kongemose for hunting and protection, rewarded with food they might otherwise have found almost impossible to acquire. A bit like today.

So consider that environmental changes – such as those driven by post-glacial sea level rise – also stressed other

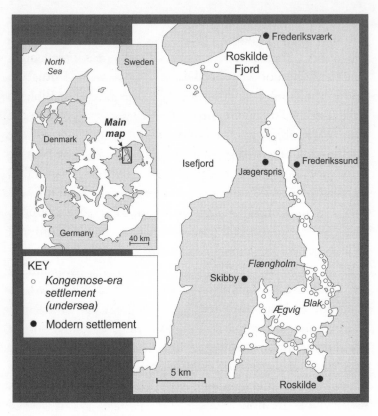

Figure 6.1 *Underwater settlements from the Kongemose era (7,200–8,000 years ago) in Roskilde Fjord (Denmark).*

creatures besides humans. And just as the Kongemose people altered their diets, switching from mostly land-based foods to seafoods as sea level drowned the lands of their ancestors, perhaps the dogs too became pragmatists. Seeing how their usual diets could not be sustained, perhaps they decided to ally themselves with humans, exchanging their independence for a social compact that would see them forever fed. I don't know about your dog, but mine nods sagely when I quiz her about this.

Two and a half months after his 600th birthday, the patriarch Noah climbed aboard his ark and firmly closed all the doors. Then 'the fountains of the Great Deep burst apart and the floodgates of heaven broke open', a situation said in the Book of Genesis (7:11) in the Christian Bible to have lasted 40 days and nights. A similar story is told in the Qur'an (11:48), where it is noted that a large part of the Earth was submerged. An even earlier version of the Noahic flood story may be in the poem, written more than 4,000 years ago, known as the *Epic of Gilgamesh* in which it is said that a man named Utnapishtim was directed to build a massive seven-storey ark in anticipation of an imminent global flood. In most versions of the story, the water subsided a few months later and Noah/Utnapishtim and his family were tasked with repopulating the Earth on which everyone else had drowned.

In almost every part of the world, traditional stories concerning the time of a great flood are found. But consider why such tales are told. Go back a few thousand years to when a memorable flood occurred. Unable to read or write, its survivors are nonetheless eager to tell of it, not least to warn their descendants that such tumultuous events do occasionally happen. But as time goes on, the details of the original flood are embellished and eventually the story becomes enshrined as myth in a people's oral history, handed down from one generation to the next. Then 500 years later, there is another memorable flood. Then another, and another. The resulting oral traditions will speak not of four floods, rarely even multiple floods, but almost inevitably of a single flood, its size amplified to stupendous proportions. Exaggeration and conflation of events are devices commonly employed in oral societies to enhance the recollection and retelling of such stories. But it does not make them literally true. A truly global flood is an impossibility.

Let us look at the practicalities. If a global flood occurred tomorrow, where would all the extra water come from? It could really come only from frozen water locked up as ice on the land. Certainly, if all that ice melted, the ocean surface would rise about 65m, which might produce a worldwide flood but certainly not submerge every inhabited place. Yet even considering the possibility of 65m of sea level rise, there is evidence, difficult to refute, that the East Antarctic ice sheet (in which almost 90 per cent of land ice is secured) has been stable for the past 35 million years. It is possible that its smaller, less stable sibling, the West Antarctic ice sheet, collapsed and melted just 120,000 years ago, but the resulting 6m sea level rise may not have affected every part of the world coastline, as explained in Chapter 13.

In none of the high-precision geological archives of past climate changes where the record of a massive global flood might be expected to show up is such an event recorded. The inescapable conclusion is that most flood myths can probably be traced back to localised events in different places at different times, memories merged and exaggerated to sustain them. For when we are not thinking as empiricists – as impartial observers of a particular phenomenon – when, in other words, we do not care for the accuracy of our observations, we may report the magnitude of that phenomenon by the degree to which it affects us personally: 'the spider in the shed was the size of my hand, I swear', when it was actually no bigger than your thumbnail. It was the same in ancient societies.

There are two comparatively well-documented situations – one from the Black Sea (Eurasia), the other from the Sunda Shelf (Southeast Asia) – where stories about a localised flood evidently persisted and were widely dispersed by its survivors across a vast region, becoming the foundation of stories about a supposedly global flood.

The first set of stories comes from the Black Sea (Figure 6.2), which was smaller than it is today during the last great ice age around 20,000 years ago, due largely to the lack of any connection to the rest of the world ocean.[4] Today this connection lies along the Bosporus Strait, a narrow shallow channel linking the Black Sea to the Mediterranean and thence to the other parts of the world ocean. A rather neat story, heartlessly disproved by recent research, is one in which the Black Sea Basin during the last ice age was home to some of the world's earliest farmers, drawn there by the region's fertile well-watered soils and benign climate to successfully experiment with crop and animal domestication long before their counterparts elsewhere in this region. Then, the story goes, as sea level started rising in the eastern Mediterranean after the end of the ice age, slowly the water level rose in the Sea of Marmara and crept along the Bosporus, until one day a small waterfall appeared along

Figure 6.2 *The Black Sea and its connection to the Mediterranean Sea.*

the south-west side of the Black Sea Basin. Soon the trickle turned to a torrent, forcing the nascent agriculturalists from their lakeside homes in a hurry.

'They fled with family, the old and the young, carrying what they could, along with the fragments of the other languages, new ideas, and new technologies gathered from around the lake'.[5]

The refugees spread across surrounding lands, eventually displacing their earlier inhabitants and leading to the development of agriculture-based societies throughout the Near East and Europe, a process referred to as the Neolithic Revolution.[6]

The stories about a great flood preserved by these refugees found their way into oral histories that became the foundations for the *Epic of Gilgamesh* and its later embodiments in stories about Noah. In turn, this led to a connection being made between these stories and the flooding of the Black Sea Basin, something that – in a regrettably premature 1999 press release from the National Geographic Society – was said to have provided 'conclusive proof that a flood of Biblical proportions inundated an area north of Turkey about 7,500 years ago – a timetable and location that virtually match the Old Testament account of Noah'.[7] This was an overstatement.

It is indeed possible that the Noah-Utnapishtim stories incorporate memories of mighty floods, driven by post-glacial sea level rise, somewhere along the shores of the eastern Mediterranean, but not from the Black Sea. Research published in 2018, based on reconstructions of ancient ocean water salinity in the eastern Mediterranean, shows that 11,000 years ago the water level in the Black Sea was already higher than that in the Bosporus, so water flowed in the other direction to that envisaged in the sequence of events just outlined.[8] This means that the Black Sea is unlikely to have been the place either from which the Neolithic Revolution was driven or that where the Noah–Utnapishtim stories originated.

Yet this does not mean that the Neolithic Revolution did *not* originate from a movement of flood-displaced migrants. Nor does it mean that the Noah–Utnapishtim stories have no possible basis in fact. It just reminds us that there are nuances in our understanding of ancient human history that are not always easily detectable. Identifying these nuances and refining our understanding as a result is exactly how the finest research often progresses.

The same research that shows the Neolithic Revolution could not have been driven by Black Sea flooding raises other intriguing questions. For instance, it shows that the spread of agriculture across Europe was twice interrupted, 8,400 and 7,600 years ago, by rapid rises of the ocean surface in the eastern Mediterranean – likely to have been caused by catastrophic events half a world away in North America.

The earlier involved the abrupt emptying of Lake Agassiz, a massive body of water covering most of modern Manitoba that formed after the melting of the thick ice sheet that covered most of Canada during the last ice age. Unable for thousands of years to find a way to the sea, this body of meltwater perched in the centre of the North American continent until one day the ice wall holding its north-east side crumbled, and the lake water gushed into Hudson Bay and thence the North Atlantic Ocean.[9] The freshwater dispersed, but some was squeezed through the sphinctic Strait of Gibraltar into the elongate Mediterranean Sea where, driving eastwards, it eventually ran up against the coast. Here it caused an almost instantaneous rise in sea level of almost 1.5m, permanently drowning coastal lands in ways that would temporarily have halted people's practice of lowland agriculture.

The later rise of sea level was far more momentous. For although the giant Laurentide Ice Sheet that had swathed

so much of North America during the coldest time of the
last great ice age (18,000–22,000 years ago) had largely
melted by 7,600 years ago, sizeable pockets persisted at this
time. It was the final decay of the Labrador section of this
ice sheet around that time that led sea level to jump along
eastern Mediterranean coasts by as much as 4.5m.[10] Three
times greater than the earlier event, we can readily believe
that such a rapid rise of sea level here could have spawned
flood stories of the kind that plausibly underpin those of
Utnapishtim and Noah. For this was not merely a case of a
single stretch of coast being temporarily flooded, as might
happen with huge waves like tsunamis. What happened
here was a large-scale inundation – the water did not recede
subsequently. Every beach, every bay, every island became
rapidly submerged, no one and no place spared; the entire
eastern Mediterranean was 'drowned'. So, for the people
living there, it must have seemed that indeed the entire
world had been submerged. The story was told to each new
generation until the point was reached where the description
of the actual event seemed so implausible that it was
exaggerated to ensure its memory endured. A regional
flood became a global flood. It reached not just the foothills
but covered all the high mountains. Everywhere was
underwater, only those on board the ark favoured by God
survived. At this, the eyes of the fireside audience grew
wider. The listeners vowed to one day tell this story to
others. And so it went on.

A second set of stories comes from the islands of Southeast
Asia that rise from the comparatively shallow Sunda Shelf,
a place that was dry land (later named Sundaland) when sea
level was lower during the last ice age. The same is true for
the neighbouring Sahul Shelf, which extended from the

edge of the north coast of Australia during ice ages, linking this continent to the island of New Guinea, as shown in Figure 6.3.

During earlier ice ages, the exposed Sunda Shelf played an important role in the dispersal of plants and animals in this part of the world, but during the last one, for the first time, modern humans were also present. There is little

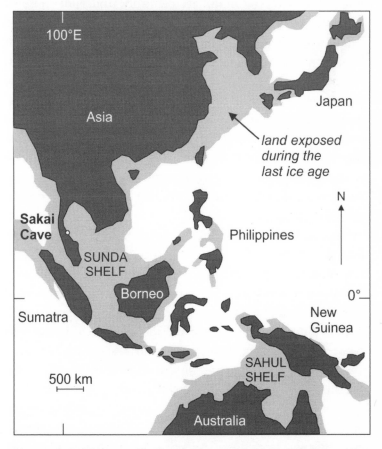

Figure 6.3 *The mix of land and sea in Southeast Asia and Australasia changed profoundly in the past 20,000 years because of a rise in sea level of 120m. Dark shading shows today's land areas, pale shading shows the additional land 20,000 years ago.*

direct information about where or how these humans lived, largely because most of the prime human habitats at the time (coastal and riverine lowlands) are now under more than 100m of water; buried beneath thick piles of volcanic ash and mud washed into the area by massive rivers like the Chao Phraya and Mekong; even grown over by coral reefs. Of course, none of this has prevented a veritable storm of speculation about 'the drowned continent of Southeast Asia', ranging from the outlandish to the credible.

Some of the former actually propose Sundaland was Plato's Atlantis, while at the other end of the spectrum lies the work of Stephen Oppenheimer who infers from linguistics, genetics and even mythology that there was a flourishing civilisation centred on the Sunda Shelf during the last ice age. Sea level rise brought an abrupt end to the development of this civilisation, forcing its people to disperse into adjacent regions.[11] Oppenheimer's argument is compelling, the evidence substantial and generally convincing, but irrefutable proof remains elusive. And probably always will.

For a long time, mainstream science held a somewhat contrary view. For while accepting that post-glacial drowning of Sundaland impacted its terrestrial inhabitants, the question of whether these included humans was once a question of some debate, would you believe? For although modern humans, including ancestral Aboriginal Australians unquestionably reached parts of Sundaland (and then crossed 70km-wide ocean gaps to Sahulland) around 70,000 years ago, the prevailing scholarly wisdom for the past few decades has been that this vast region remained only scantily peopled until about 5,500 years ago when it was overrun by waves of rice farmers from the north, a process akin to the Neolithic Revolution in Europe. For many scientists, this scenario confirmed the earliest site of rice domestication as having been in China as well as demonstrating why Austronesian ('out-of-Taiwan') languages are spoken throughout South-east Asia today.

But as is increasingly the case, the movements of ancient peoples reconstructed using their DNA tell a different story. Recent studies show that Sundaland was occupied continuously by modern humans since the ancestors of Indigenous Australians arrived there more than 70,000 years ago and, in a lethal blow to the orthodoxy, that there was a major *outflow* of modern humans from Sundaland into East Asia (eastern China, Japan, Taiwan) around 12,000 years ago, plausibly because of the spreading effects of post-glacial sea level rise.[12]

It may be that the now-submerged Sunda civilisation was more advanced than others at the same time in South and East Asia because the land it occupied was uncommonly productive. Its inhabitants occupied a region of fertile lowlands built from alluvium carried down from the upper catchments of large rivers. The principal crop that sustains the modern inhabitants of hot, wet Southeast Asia today is rice, and some contend it was rice that sustained their ancestors in Sundaland more than ten millennia ago. The problem with proving this is obvious. How can you discover what people harvested and consumed in a place 10,000 years ago when that place is now somewhere under tens of metres of water? We have no choice but to resort to inference.

Now drowned, the riverine lowlands of the Sunda Shelf would have lent themselves well to rice cultivation, and it is possible this started here before the region was inundated and then spread (or diffused) outwards as the ocean surface gradually rose. There are gobbets of evidence to support this scenario, including the discovery of rice grains on 9,400-year-old pottery from Sakai Cave in southern Thailand (location in Figure 6.3), just above one of the broadest areas of the floor of the Sunda Shelf. Somewhat more compelling is the linguistic evidence for rice cultivation apparently beginning in the lower Red River Valley (Vietnam) and thence diffusing up-valley into China and India. Yet ranged against these possibilities is a formidable body of empirical evidence showing that rice was domesticated first around 10,000 years ago, maybe earlier, in the lower Yangtze Valley (China), although wild varieties could have been sustaining people here and elsewhere far longer. There is, of course, no reason why rice cultivation could not have begun independently in both Sundaland and the Yangtze Valley.

You may regularly consume cucurbits – from cucumbers to pumpkins and melons. In the same family are the bottle gourds (*Lagenaria*), the shells of which, when emptied, can be dried and used as water containers. It is not too long a shot to suppose that these bottle gourds, which originated in tropical Africa, were carried intentionally by the earliest groups of modern humans as they spread across the prehistoric world, not least to carry water but also as musical instruments, fishing floats and penis sheaths.[13] Almost certainly, the bottle gourds were part of the armoury of domesticated plants that helped sustain the people of Sundaland and accompanied them on their subsequent journeys of dispersal.

An association between bottle gourds and rice is common in many Southeast Asian cultures and their mythologies, suggesting that rice may indeed have been growing in now-undersea Sundaland. For example, bottle gourds are today regarded as strengtheners of rice (nicknamed 'sweethearts'), explaining why in paddy baskets – which contain rice sheaves for future planting – gourds are also often placed. In the numerous Southeast Asian flood myths, some of which may plausibly include echoes of Sundaland submergence, the gourd is the most common vehicle for human survival. Consider the origin story of the Kharian people from Orissa in India, who tell that God gave a gourd seed to two siblings who had survived a deluge and instructed them to plant it. On the vine of the gourd appeared three kinds of grain – millet, dry rice and wet rice – which sustain the Kharian to this day.[14]

The point here is that, in the absence of clear evidence, we can infer that the people who occupied now-drowned Sundaland were driven into northern lands by rising sea level, carrying with them words and practices and materials that influenced those of the peoples they encountered, laying the foundations for the rich admixture of cultures that is Southeast Asia today.

Some people from drowning Sundaland may also have sailed east to the Pacific where, as we saw in the last chapter, there are many island peoples who claim a land named Hawaiki as their homeland, a place from which their ancestors had to move when it was submerged. Māori scholar Te Rangi Hīroa (Sir Peter Buck) referred to Hawaiki as a land below the sea, while it was also considered a 'veritable Hades, the shadowy Under-world of death, and even of extinction'.[15] So consistent and so ubiquitous are such stories in Pacific Island cultures that it is possible they echo – in a similar way to stories about Noah–Utnapishtim in Europe and the Near East – ancient memories of land submergence. As happened to Sundaland.

Biogeographers – those people preoccupied with explaining why particular species of plants and animals are found where they are – have some uses to other scientists. One use – discussed here – is to explain how particular species of terrestrial plants and animals, unable to swim or fly, apparently managed to cross vast stretches of unbroken ocean to reach islands far away. One explanation for this is that they first colonised these islands when the ocean surface was lower, as it was during past ice ages, when other islands (now submerged) existed. These islands acted as stepping stones, allowing terrestrial (and shallow ocean-dwelling) plants and animals to break their journeys, otherwise impossible, on dry land.

A good example is provided by the Pitcairn Island group of the eastern Pacific. Pitcairn is comparatively small, you can walk around it in a day. It is also comparatively remote, a providential attribute explaining why the *Bounty* mutineers went undiscovered here for 25 years before the long arm of British justice finally caught up with them.[16]

Even though Pitcairn was unoccupied when the *Bounty* mutineers landed there in 1789, hundreds of years earlier it had been home to Pacific people who also occupied, sometimes only seasonally, the other islands in the group – Ducie, Henderson and Oeno (Figure 6.4). Of these, two are today atolls – low, resource-poor islands – while Henderson is a raised limestone island that supported a permanent population for around 600 years until it was finally abandoned around AD 1450.[17] The islands of the Pitcairn group are hundreds of kilometres apart, perhaps no real challenge to the accomplished seafarers who once lived there, but a significant barrier to the movement of land plants and flightless animals, particularly in earlier (inter-glacial) times when sea level was even slightly higher than at present and Ducie and Oeno were wholly underwater.

Given all this, it was astonishing to discover that the native plants and animals found today on all four islands in the Pitcairn group, which lived there long before any humans arrived, show clear affinities with one other. Together with the island groups to the west (like Mangareva and the Tuamotus), they comprise a distinct biogeographic province within which species of terrestrial plants and animals have moved back and forth for millions of years.[18] So the question arises, how were these organisms able to repeatedly cross ocean distances of nearly 400km between these islands? And also, given that low Ducie and Oeno were submerged every 100,000 years or so by high (interglacial) sea level causing their terrestrial biotas to be wiped out, how is it possible for these to have been subsequently reinstated? As though nothing had happened.

Well, it is not helpful to appear too dogmatic about such matters because other explanations may one day be forthcoming, but at the moment it seems the answer lies with the presence of now-drowned islands in the region that were emergent (dry and high) during every ice age when sea level was much lower. These insular stepping stones, their locations illustrated by crosses in Figure 6.4, enabled the diffusion of plants and animals within the Pitcairn group time and again.

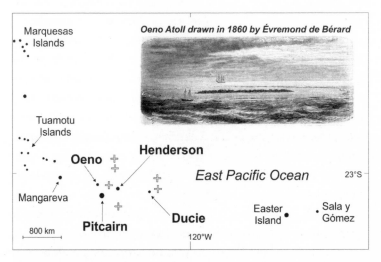

Figure 6.4 *Some of the remotest specks of land in the vast Pacific Ocean were far less remote 20,000 years ago. Then, because sea level was 120m lower, other islands (locations shown by crosses) existed. In the Pitcairn group (including Ducie, Henderson and Oeno), there were six extra islands that aided dispersal of land plants and animals within it.*

To understand how this worked, imagine a time during the last inter-glacial period about 125,000 years ago when the only dry land in this part of the Pacific was lofty volcanic Pitcairn Island and far less lofty limestone Henderson Island. On Pitcairn grew a small tree, one of the spurge family known only by its scientific name *Glochidion pitcairnense*, the seeds of which were dispersed by birds like the bristle-thighed curlew that still today flies between these islands. This explains why we find this spurge growing today on both Pitcairn and Henderson. When sea level was lower during the last ice age, curlews would also have carried seeds of this rare plant to other now-submerged islands as well as Ducie and Oeno, where they germinated on what would then have been high limestone islands.

A plant found on all four islands in the Pitcairn group today is a flowering shrub of the borage family, common along low windswept coasts on Pacific islands, named the octopus bush (or *Heliotropium foertherianum*). It is prized by the islanders for its medicinal properties – particularly to treat fish poisoning – and as firewood. Its seeds are dispersed in water, carried by ocean-surface currents between islands. For both the spurge and the borage, you can see how the presence of island stepping stones would have aided their dispersal among the islands of this spread-out island group.[19]

It is the absence of any similar island stepping stones between Pitcairn and Easter Island, 2,200km away to the east, which explains why its land biota is so different to that of the Pitcairn group. Easter Island is a biogeographic enigma. Famously, Thor Heyerdahl argued its plants came from South America and had been carried to the island by its earliest human colonisers who then went on to colonise most other Pacific Islands. Heyerdahl was wrong on both counts. The defining elements of the Easter Island biota have survived through the alternating emergence and submergence of an island chain at this location for at least 30 million years. The first human colonists of Easter Island arrived only within the past 1,500 years and came from the islands to the west, not east from South America.[20]

A kindred enigma is why 'drowned' coral reefs exist in some parts of the world's ocean. After all, we know that today most corals live close to the sea surface, centrepieces of vibrant, colourful and diverse ecosystems that need sunlight to survive. So why should any such coral reef ever drown? How did some become submerged hundreds of metres below the sunlit ocean surface, in places so dark and cold that you might wonder how anyone ever found them there?

Coral reefs are extraordinary things. The Great Barrier Reef off the east coast of Australia is the only organic structure on Earth visible from the moon. Like other reefs, it has a veneer of living corals growing just below low-tide level on the piled-up remains of their dead ancestors. Found only in warmer oceans, coral reefs are also ecologically somewhat remarkable. Made from numerous individuals, reefs appear to respond communally to external stresses. And over the past 450 million years, coral reefs have shown themselves to be remarkably resilient, weathering massive swings in climate and sea level to flaunt their sublime beauty and share their riches with us today.

By definition, coral reefs must grow close to low-tide level in order for the photosynthetic algae that their polyps host to receive enough sunlight to produce the food that both they and their hosts need to live. So, when sunlight is reduced, perhaps because the sea surface rises or because the surface water becomes more turbid, crisis looms. If sea level rises above a coral reef for a prolonged period, several decades perhaps, the species composition of its living veneer will eventually change to enable the reef surface to grow upwards, keeping pace with the rising sea level that might otherwise submerge it. But there is not much to be done about increased surface-water turbidity, something increasing off the mined and overgrazed hinterland of north-east Australia adjoining its Great Barrier Reef.

Turbidity is but one of three major constraints that the natural environment poses for corals. Another is salinity. Where ocean-surface water is too fresh (not salty enough), as is the case where large rivers enter the sea, corals cannot grow and reefs will not develop. A third constraint is temperature. Corals are fussy organisms that often die when the temperature of the water in which they are living falls below 20°C or rises above 30°C for prolonged periods.[21] It seems likely that some unmanageable combination of changes in sea level, turbidity, salinity and temperature was responsible for drowning some coral reefs.[22]

Unwelcome human impacts on some of the world's coral reefs are approaching a crisis point. Rising sea level and temperature are currently the most common culprits although the acidification and deoxygenation of ocean water are likely to join them soon. In some scenarios, those corals currently living at (or just below) the ocean surface are facing extinction or at best significant losses. If such a scenario eventuates by, say, the year 2050, then mesophotic coral reefs, which live at depths of 20–30m, may become important refuges for corals and allow these remarkable organisms to survive future warming.

In addition to shallow drowned reefs there are ones that exist, devoid of the lifeforms associated with their shallower-water counterparts, at depths in excess of a kilometre below sea level. Such reefs drowned through a combination of land sinking and sea level rise. To put this in perspective, consider that the continuing subsidence of the largest youngest islands in the Hawaii group, a classic location for drowned reefs, occurs at rates slow enough for reefs to be able to grow upwards at the same rate. But when island sinking coincides with sea level rise, many reefs simply cannot cope and become drowned. All around slowly sinking islands like Hawai'i (the Big Island), O'ahu and Lana'i, there are underwater staircases of drowned coral reefs, each marking a time when post-glacial sea level rise amplified submergence and led to reefs becoming drowned. One of the most deeply drowned reefs lies some 1,150m below sea level on the side of Māhukona volcano, an underwater volcano off the north-west coast of Hawai'i Island.[23] Another dead reef was drowned by a burst of rapid sea level rise just after the end of the last ice age. Having been alive around 15,000 years ago, it is today found in several places off the coast of the island of Hawai'i at a depth of 150m.[24]

On the morning of 28 August 1970, the scientific research vessel *Glomar Challenger* was probing the floor of the Mediterranean Sea south of the Balearic Islands. Beneath almost 3km of ocean, the drill was pushing its way down through the soft sediments that you expect to find just below the ocean floor, when abruptly and unexpectedly the drilling rate slowed. A hard layer had been encountered. The next day the first samples from this hard layer were brought on board and were revealed as anhydrite, a type of rock (evaporite) formed by relentless evaporation in hot deserts.[25] The inescapable conclusion was that at the time this anhydrite formed on the floor of the Mediterranean Sea, it had been dry land – a hot desert similar to those found today in countries like Egypt and Libya along its southern border.

Subsequent ocean-floor research in the Mediterranean revealed not only the vast extent of this particular evaporite layer, signalling that almost the entire Mediterranean had once been desert, but also the presence of other evaporite layers sandwiched between the more pliant ocean-floor oozes. The Mediterranean had evidently been dry land on several occasions. Dating shows that the greatest of these occurred during the Messinian stage of the late Miocene epoch, 5.3–7.2 million years ago, when global sea level fell and rose massively.

The Mediterranean Sea occupies a basin within the Earth's crust; its essential form is that of a bathtub. Emergent continental margins comprise three of its sides while the fourth – where the 13km-wide Strait of Gibraltar lies – has for much of the history of the Mediterranean alternately opened and closed, respectively permitting or shutting off the exchange of water with the Atlantic Ocean. During the period of falling sea level at the start of the Messinian, the tap shut off and the Mediterranean Sea became a giant lake. As with any lake in a place where evaporation persistently exceeds precipitation, the shallower water around its edges began evaporating, the salts released from it accumulating in great corrugated expanses of soft sediment named evaporites. So long as these evaporites remained damp, they stayed soft, but in the driest parts of the ancient Mediterranean, where every drop of water was boiled out of them by the relentless Messinian sun, they hardened to form the anhydrite that the *Glomar Challenger* first brought to human view 50 years ago.

During the Messinian, the water level in the Mediterranean dropped an astonishing 2,500m. The water evaporated from it ended up through precipitation in the other oceans of the world. Since this water was essentially salt-free, the salinity of the world's oceans dropped 6 per cent as a result, which had effects on ocean chemistry, ocean circulation and even on world climate. This is why this 630,000-year-long event is referred to as the Messinian Salinity Crisis.

As the Crisis progressed, the giant Mediterranean lake became smaller but never apparently so small that it allowed a broad land bridge to emerge between North Africa and Southern Europe. Yet animals from both sides moved into the centre of the basin, following the receding water line, to occupy lands that later became isolated islands within the Mediterranean, stranding these creatures and driving them down evolutionary cul-de-sacs. Among the biological rarities resulting from this are the antelopes, elephants and hippopotami from North Africa, which reached what are now the Mediterranean islands of Crete, Cyprus, Malta and Sicily. These species, all now extinct, underwent insular dwarfism, a phenomenon resulting from isolation and the lack of any need (to dissuade other creatures from attacking you) to remain large-bodied. The adult Cyprus Dwarf Hippopotamus (*Phanourios minutus*), for example, reached about 76cm in height.

The end of the Messinian Salinity Crisis about 5.3 million years ago began when the tap from the Atlantic came on again. There was a rapid flooding of the Mediterranean Basin, transforming it from a hot dry region surrounding a large island-dotted freshwater lake to an ocean, connected once again to the rest. New species of sea creatures, which had evolved in warm yet salt-reduced Atlantic waters during the crisis, flooded in while countless freshwater species living in the Mediterranean became extinct because they could not adapt fast enough to the rapidly changing environment. Vast swathes of land, much plastered with evaporites, were drowned. The soft varieties quickly dissolved, but the hard anhydrites proved insoluble and the salt they contain is still lost to the world's ocean. Even today, more than 5 million years after the Messinian

Salinity Crisis ended, the waters of the Mediterranean are less salty than those of other oceans for this reason.

Most of the floor of the Messinian Mediterranean now lies buried beneath tens of metres of sediments below hundreds, even thousands, of metres of ocean. A world in shadow whose character is impossible to reconstruct in detail, but one which we know once existed.

CHAPTER SEVEN

Deep in Shadow: Ancient Lands
Now Hidden

On 20 February 1835, eager to be on dry land after months at sea, 27-year-old Charles Darwin went ashore near Valdivia on the west coast of South America. Walking through the woods, he lay down to rest. The earthquake, he reported, 'came on suddenly, and lasted two minutes, but the time appeared much longer'. Darwin scrambled to his feet,

> ... but the motion made me almost giddy: it was something like the movement of a vessel in a little cross-ripple, or still more like that felt by a person skating over thin ice, which bends under the weight of his body. A bad earthquake at once destroys our oldest associations: the earth, the very emblem of solidity, has moved beneath our feet like a thin crust over a fluid; one second of time has created in the mind a strange idea of insecurity, which hours of reflection would not have produced.

When he found time to reflect on the effects of this earthquake, which he observed with increasing despondency over the following two months, Darwin was able to reconcile them with the ideas of Charles Lyell, whose three-volume *Principles of Geology*, had been published shortly before Darwin had left England in December 1831 and was something he read on what was to be an almost five-year circumnavigation of the Earth on HMS *Beagle*. Darwin concluded:

... the frequent quakings of the earth on this line of
coast are caused by the rending of the strata, necessarily
consequent on the tension of the land when upraised,
and their injection by fluidified rock.[1]

Lyell had argued that continents were raised parts of the
Earth's crust, pushed upwards during successive earthquakes
(among other processes), something that Darwin's
observations in Valdivia and Concepcón convinced him to
be correct. Not only had he witnessed the ground shaking,
but also its elevation during the 1835 earthquake. During
his travels through the hinterland of the Pacific coast of
South America, Darwin saw beds of seashells hundreds of
metres above the coastline, demonstrating to his satisfaction
that earthquakes had indeed elevated it on numerous
occasions in the past and that Lyell's core belief – that
continents were upraised parts of the crust – must therefore
be correct.

Yet Darwin went a step further, extending Lyell's
arguments about land uplift by proposing that the ocean
basins were areas of complementary sinking (subsidence).
Darwin based his opinion on the existence of atolls,
remarkably before he had ever seen a 'true coral reef',
deducing correctly that they formed as a result of coral reef
growing upwards from the flanks of a subsiding underwater
island. After the Beagle left the coast of South America to
sail west across the vast expanses of the Pacific and Indian
oceans, Darwin was also able to visit several oceanic islands
and see for himself how different the rocks forming them
were from those of the continents. From this he concluded
that continents and ocean basins were not only moving
vertically in different directions but were also quite distinct
in composition, an inference that laid the foundations for
his prescient division of the Earth's crust into continental
and oceanic types.

Inevitably, the actual situation is more complex than envisaged by Lyell and Darwin. Earthquakes can elevate coastlines, but can also drop them; the Chilean coast is indeed an excellent example of the former. Continents are not by definition raised pieces of the Earth's crust any more than all parts of the ocean 'basins are sunken. Their difference in level is explained by their relative densities; lighter continental crust 'floats' on denser oceanic crust, just as ice floats in a glass of water. Yet the fundamental distinction between these two types of crust remains central to ideas about Earth-surface evolution, a tribute to the keen observations and well-honed minds of pioneer thinkers like Lyell and Darwin.

The distribution of continental and oceanic crust on the Earth's surface is not as neat as a purist might wish, meaning that surprises, providing key insights into a particular period of our planet's evolution, sometimes pop up where you would least expect. There are tens of thousands of oceanic islands. Many are cloaked in dense rainforest so it is sometimes almost impossible to find any exposures of the underlying rocks. So, again, geologists are beholden to biogeographers for identifying plants on particular islands that appear literally out of place, specifically the occurrence of 'continental' plants growing on 'oceanic' islands. The rationale is that the continents have been in existence far longer than oceanic islands (almost 5,000 million compared with 80 million years) so the vegetation native to many continents is quite different, typically more diverse and with longer lineages, than is found on such islands.[2]

You might not think there were many island groups more oceanic than that of Tonga in the central South Pacific Ocean. And in the south of Tonga lies the island named 'Eua, an island made almost entirely from raised coral-reef limestones. The western side of 'Eua where most people live is comparatively low, but as you walk east through dense forest, the land rises sharply in a series of giant steps – the fronts of ancient uplifted coral reefs – until you reach the summit ridge 300m above sea level, which then drops sheer to the tiny, almost inaccessible, beaches below. The forests of 'Eua include an endemic conifer (*Podocarpus pallidus*), the presence of which must have come as a huge shock to the botanists who first identified it. For podocarps are species of tree that dominate ancient Gondwanan forests that developed not on islands in the middle of oceans but in the oldest parts of the southern continents. Podocarp forests are widespread today in New Zealand and South America, but were once even more extensive, growing throughout Antarctica before it became covered with ice.

Probably the first geologist to visit 'Eua was Joseph Jackson Lister who landed there in 1888. Along the island's steep east coast at the foot of sheer 200m-high cliffs, Lister saw to his amazement that the limestones had, in a few places, been eroded to expose the volcanic nucleus of the island across which they had been plastered.[3] Later research showed that, far from being the youthful volcanic rocks you would expect to find on islands in the central Pacific, the lowermost outcrops of volcanic rocks on 'Eua formed tens of millions of years earlier and are of continental (not oceanic) origin, a discovery that shocked the community of oceanic geologists and clearly demanded a unique explanation.

This came in the form of the 'biotic shuttle hypothesis', the idea that in the distant past as continents broke apart, small fragments of these – carrying distinctive continental plants and animals – drifted across the ocean basins, periodically colliding with oceanic islands and allowing some of those plants and animals to disembark.[4] This explains why ancient continental podocarp species grow today on 'Eua but it also seems likely that, from the presence of ancient volcanic rocks poking out from beneath the limestones along the back of this island's east coast beaches, 'Eua itself may be just such a continental fragment.

Stranded in the central Pacific, its true colours fortuitously revealed, we now know that this 87km² island was once part of the continental island of New Caledonia, which was formerly joined to Australia and New Zealand. The three parted company around 85 million years ago, a process driven by the slow opening of the Tasman Sea, but were earlier part of a massive southern continent – an amalgam of most of today's southern continents – known as Gondwana. Gondwana itself was formerly joined to Laurasia, an amalgam of the world's northern continents, and about 220 million years ago they all formed a supercontinent that today we call Pangaea.

Pangaea incorporated every piece of continental crust on the Earth's surface. A true supercontinent. Pangaea was surrounded by a super-sized ocean, posthumously named Panthalassa, that was twice the size of the modern Pacific, its direct descendant. Pangaea is no more because it broke up – disaggregated – but it is still possible to see the evidence that it once existed. The simplest evidence is the shape of the continents. If you look at a map of the Atlantic Ocean, you can see how the bulge that is mostly Brazil fits snugly into the concave coastline of equatorial Africa. As it did 180 million years ago before the Atlantic began opening.

The earliest of the Earth-surface mobility models implicit in these ideas was that of 'continental drift', the name given to the process by which Pangaea disaggregated. Early explanations for continental drift, a theory resisted for decades by most geologists, pictured the continents as the movers, ploughing – like massive icebreakers – through the more yielding oceanic crust. This idea was wrong. For while Earth-surface mobility is unquestioned by geologists today, it is now clear that the continents are passive riders on the mobile oceanic crust, which in turn sits atop a layer of semi-molten rock named the asthenosphere more than 80km beneath our feet. The rock in the asthenosphere moves like treacle, it is viscous and also extremely hot, at least 1,200°C (2,372°F). In some places, usually from sea-floor sutures running along the axes of ocean basins, semi-liquid rock welling up from the asthenosphere slices open the oceanic crust, forcing the two sides apart and creating new crust. This process is balanced, more or less, by the thrusting downwards of old oceanic crust along deep ocean trenches, a process that results in the destruction of this crust and its eventual recycling through the asthenosphere. In this way, the entire Earth's crust can be pictured as divided into a series of rigid interlocking 'plates' that are constantly moving. Most earthquakes and volcanic activity are attributable to these movements.

The current model of Earth-surface mobility is called 'plate tectonics' and it has been successfully used to explain the alternating aggregation and disaggregation of continental crust into supercontinents. For a long time, scientists believed Pangaea to have been the only supercontinent in the Earth's history, but it is now clear that around 1 billion years ago Earth's continental fragments had also aggregated into a supercontinent that became named Rodinia. Being so ancient, the evidence for the existence of Rodinia is far less glaring than that which allowed Pangaea to be recognised.

The story begins in the early 1990s, when geologists studying earlier and more obscure times in our planet's history found there had been an unusual concentration of mountain building between 1–1.3 billion years ago. These mountain ranges comprised rocks that had been folded and crushed, a result of the Earth's crust being squashed when two adjacent plates collided. It was deduced that this period of time had been marked by an unusually high number of continental collisions, exactly as would have occurred at a time of supercontinent aggregation. So, we learned of the existence of Rodinia.

Rodinia existed for 350 million years until it began to break up as a result of the build-up of heat beneath it. Beginning around 750 million years ago, the break-up of Rodinia led to a massive increase in the diversity of Earth-surface environments. Think about it. If all the dry land on Earth was today in a single landmass, then its environments would lack diversity; coasts and mountains and endless dreary inland plains. Some parts warmer and some cooler, but with zonal temperature contrasts subdued by the topographic regularity. Rather like a massive version of modern Australia, the supercontinent of Rodinia would have had moist coasts, but most of its land area would have been uniformly dry. Not an especially inviting prospect for most lifeforms.

A global rise of sea level accompanied the break-up of Rodinia, producing lots more coastline and increasing climatic diversity, which then increased environmental diversity. New exciting opportunities appeared for organic life, allowing it to evolve and occupy newly created niches, especially along coastlines, both above and below the ocean surface. All this resulted in a massive increase in biodiversity, the so-called Cambrian explosion of life. The earliest animals appeared some 540 million years ago, the first fish 500 million years ago, the first land plants about 420 million years ago. All, it is considered, as a consequence of the break-up of Rodinia.

Tantalisingly, hints have been detected in the geological record of supercontinents even earlier than Rodinia. Details are spare, contentious, yet sufficient to convince many that the Earth's crust has passed through a series of supercontinent cycles. So far so good, but as you might guess, all this talk of supercontinents stirred the paludal imaginations of pseudoscientists who, contemptuous of time depth and the scientific method, saw in this a grand opportunity to rewrite human and planetary pasts.

For more than 100 years, pseudoscientists have lined up to tell us that there is a vast 'sunken continent' in the centre of the Pacific. There is not. No question about that. But to my knowledge, none of these people have acknowledged the one sunken continent that did indeed once exist in the Pacific, but which is no longer there (spoiler alert: it didn't sink, it moved and hid). This is understandable, given that little in English has been written about it and all of this has appeared only in professional geological journals. This ancient continent – it has no name – no longer exists in the Pacific. All trace of it here has gone, and we know only by inference that it was ever there. But to a geologist, the evidence is glaring.

Rocks fall off steep cliffs, often forming slopes or fans of talus – loose material that piles up at the foot of a cliff. Where the cliff foot is deep underwater, the rocks that fall off still accumulate in distinctive fans of talus but, because of the water through which they drop, they are generally better sorted and extend further away from the cliff foot than their counterparts on land. What I mean by sorting is that the larger (or coarser) particles of fallen rock settle closer to the cliff while the smaller (or finer) particles drift further away. So, in such undersea talus deposits, named turbidites, there is usually a gradation in particle size – from coarse to fine – away from the cliff. The importance of all this is that, by studying the nature of the sorting within an ancient turbidite, the direction from which it came can be identified.

In the Pacific Ocean off the west coast of South America, from Peru in the north to Chile in the south, ancient cemented turbidites like the 2,700m-thick Zorritas Formation are found. This is exactly what you would expect along such a high (and old) continental margin where for hundreds of millions of years large rivers rising in the Andes mountains have carried huge quantities of sediment out to sea, dropping much of it on the ocean floor at the foot of the cliff marking the limit of the narrow continental shelf. But the strange thing about these particular turbidites is that their sorting characteristics are the opposite of what would be expected had they originated on the west coast of South America. These turbidites coarsen from east to west (not west to east) signalling the fact that they originated along the east-facing coast of a large landmass in what is now the south-east Pacific Ocean. That this landmass was of continental origin is clear from the minerals comprising the turbidites. In the minds of some of the earliest geologists to conduct systematic surveys of the west coast of South America, these turbidites were the principal evidence for supposing that offshore there had existed here – slightly more than 400 million years ago – an ancient Pacific continent – or *un ancien continente Pacifique* in the words of Carl Burckhardt in 1902.[5]

More evidence of this ancient continent has been gathered since Burckhardt's time, notably from the presence of belts of folded rocks along the west coast of South America and the truncation of some of the north-west to south-east trending structures found there. The folded belts suggest that a large landmass once repeatedly bumped into South America, crumpling the rocks in the contact zone. The truncation of the structures can be explained as a result of that same landmass slowly grinding its way north, knocking off the ends of any structures protruding in its path.

The analogy with a ship tied up loosely to a dock, periodically hitting it, then being dragged alongside it scraping off barnacles, is almost irresistible. But remember in this case we are dealing with a huge landmass that bumped and scraped for tens of millions of years against another huge landmass. Some 'barnacles' were the size of Lima or Santiago. And the 'ship' is no longer there.

So where did the ancient continent in the south-east Pacific come from and where did it go? We can answer both questions, thanks largely to research carried out on the break-up of Rodinia, the supercontinent that existed 1 billion years ago. When Rodinia was fully aggregated, what is now the continental core of North America (called Laurentia) was joined to parts of what are now East Antarctica and Australia. As Rodinia began breaking up, Laurentia moved to a position off the west coast of South America, the time when the 'reversed' turbidites were created.

Then Laurentia moved in. Colliding repeatedly with the western margin of South America, it caused the sediments in the contact zone (including some of the turbidites) to be folded and elevated. But the eastward movements of Laurentia were relatively unimportant. Most of the continent's movement was northward and, at those times when it was in contact with South America, it ripped off the protruding edges of structures it encountered. Eventually Laurentia moved north of South America and came to rest in approximately its present position. Today we see Laurentia, the ancient Pacific continent, in much of North America, centred on Manitoba. While its shape has changed, and many other bodies of land have become stuck to it, the central part (or craton) of North America is still recognisable as that Pacific continent which once lay off the west coast of South America. Hidden but decidedly not lost.

The Pacific Ocean has existed as a discrete entity since the time Rodinia began to break up, around 750 million years ago. But in all that time, only one sizeable continental fragment ever existed within it – that just described. There have been a few smaller continental fragments that have darted their way across the Pacific, eventually becoming stuck on to its continental rim or stranded like 'Eua in its centre, but none as large as Laurentia was 400 million years ago.

Since we instinctively regard continents as big, the idea of a microcontinent might sound oxymoronic. Yet within the ocean basins, there are smallish fragments of continental crust. To understand how they got there, you might take a slice of bread, place it on a flat surface and pull it apart. However slowly you do this, there will be crumbs left behind, which is a fair analogy of why in most ocean basins today we sometimes find microcontinents far from the parent continents to which they were once attached.

Reassuringly for those of us keen to expose the manifest stupidities of pseudoscientists and new-agers trying to convince people that human evolution involved a prolonged residence on now-sunken continents, most of these continental fragments are small and were submerged long before modern humans appeared on the Earth. On the other side of the coin, it has to be admitted that there are today far more microcontinents than scientists once believed there to be. Darwin's division between continental and oceanic crust is not actually as neat as he envisaged.

Unsurprisingly, hardly any microcontinents are found in the Pacific, the world's largest ocean, because it is the remains of Panthalassa, the world ocean that surrounded the supercontinent of Pangaea. Pangaea was torn apart, creating all the other ocean basins and it is in these we find the 'crumbs' left behind by this process. But the Pacific is essentially crumbless except for tiny anomalies like that of 'Eua that arrived there in singular ways. To illustrate the

fascinating history of the genuine oceanic breadcrumbs –
sorry, microcontinents – the origins of the islands of Jan
Mayen (North Atlantic) and the Seychelles (Indian Ocean)
are described.

Today a Norwegian outpost well inside the Arctic Circle,
the Vikings may have been the first people to reach Jan
Mayen, naming it Svalbarô or 'cold coast'.[6] Today a
designated Important Bird Area, a French expedition that
landed on uninhabited Jan Mayen in 1892 had to elbow its
way through flocks of fulmars, auks and guillemots in
order to get beyond its shore, so abundant and so naïve
the island's avifauna.[7] Jan Mayen is also home to the
northernmost active land volcano on our planet, Beerenberg,
a typical oceanic volcano responsible for building the entire
island within the last 700,000 years. So far then, no
surprises – Jan Mayen seems a typical oceanic island.

Then in the 1990s, ocean-floor drilling in the area south
of Jan Mayen revealed the presence of anomalously thick,
light (low-density) crust that was deduced to be a continental
fragment sliced off the continental margin of eastern
Greenland 55 million years earlier. It became obvious that
the geology of Jan Mayen Island itself was a hopeless guide
to that of the surrounding crust, the island simply being a
younger feature formed when lava punched its way through
the comparatively thin continental crust at the northern
end of the microcontinent.[8]

On the other side of this part of the North Atlantic lies
the undersea Hovgård Ridge, a similar sliver of crust rifted
off nearby Svalbard about the same time as the Jan Mayen
microcontinent. For more than 20 million years, the Hovgård
Ridge was dry land, covered by 'dark conifer forests with
low-lying ferns … a mysterious and gloomy place' that put
the scientists who studied it in mind of Plato's Atlantis. This,
of course, is mere wishful thinking, but we do now know
that, lurking beneath the ocean surface in the northernmost

Atlantic, there are microcontinents, some once emergent, all now submerged.

Birds also feature in our second example of an ocean-bound microcontinent, the Seychelles archipelago in the western Indian Ocean. Until people first arrived on these islands in 1770, their extraordinary ecosystems, which include 15 endemic bird species, had thrived in isolation. Rather than people themselves, it is the rats, accidentally introduced, which have had the greatest impacts on the native avifauna, sparking aggressive rat eradication programmes and the establishment of rat-free bird reserves. Yet it is the islands' geology and the diversity of landscapes this created that fundamentally explain the uniqueness of the biota of the Seychelles.

The earliest identification of non-oceanic rocks, notably granite, on the Seychelles has sometimes been carelessly attributed to Charles Darwin, but the truth is he never visited these islands on his five-year circumnavigation in HMS *Beagle*.[9] Granites can be considered the archetypal continental rock – every continent has them – so their presence on mid-ocean islands is cause for some wonder.

The granitic islands of the Seychelles rise from a microcontinent, the northern part of the Mascarene Plateau, which is considered a fragment of the ancient continent of Gondwana (the southern half of Pangaea) that broke off the Indian subcontinent as this was moving northwards away from Africa 63 million years ago.[10] This part of the Mascarene Plateau is a composite geologic feature, the Gondwanan rocks in the north having become interleaved with basalts from the eruptions of the Deccan Traps in India around the same time.[11]

Browsing bookshop and library shelves, it is possible to find some accounts of 'sunken continents' or 'lost lands' in the world's oceans but, in my experience, most of these accounts are misleading, intentionally so in most cases. The writers of such fictions represent them as real and, craving scientific approbation, often fasten on the slightest hint of supporting data, sometimes mashing credible scientific information into a pulp to support their inane arguments, even dropping in an inappropriate citation to a reputable scientific source in support of these. Many of us have been duped.

Such accounts often fasten on the incompleteness of scientific studies of the ocean floor, something translated by pseudoscientists into mystery. Mystery, of course, merely means unknown not incomprehensible, but such writers invariably represent it otherwise: *The Unknown* where *Mysteries* abound. The truth is less fantastic. While scientists may not have explored every square metre of the ocean floor, there is little mystery about what is there. Imagine you have a back garden of 30m². You may dig a few holes here and there to plant fruit trees, but would you really expect to find anything wildly different by digging elsewhere?

With which we turn to LIPs – large igneous provinces – and the possibility these might qualify as ancient sunken lands. Periodically in the history of the Earth, its crust is wrenched apart and liquid rock from its interior wells up and spills out across large areas of its surface, sometimes for millions of years. The results are what we now recognise as LIPs. The existence of these have been known for decades, but some of the earliest geoscientists to study them simply could not conceive – largely on account of their enormous extent – that each was the product of a single event. A clear case of not seeing the wood for the trees.

Our focus here is on oceanic LIPs rather than those – like the Deccan Traps – exposed mostly on continents. The reason for this bias is that we are interested in understanding whether there is any possibility of there being submerged continent-sized landmasses in the middle of the oceans, as many pseudoscientists assert.

When huge masses of magma accumulate just beneath the Earth's crust, they make its surface swell – like giant pustules. As more magma pours in from the Earth's interior, the pustules grow until their crustal skin can no longer contain the magma. This then spills out to form vast, beyond island-sized, masses of new crust (LIPs), most of which now lie well below the ocean surface. Most oceanic LIPs have never been dry land. The Kerguelen Plateau, a sizeable LIP in the southern Indian Ocean, is an exception for it was actually above sea level for the first half of its 115 million years of existence.

LIPs can form anywhere on the Earth's surface, along plate boundaries or in the more placid centres of plates, on continents or on ocean floors. The conspicuous lack of correspondence between Earth's crustal plates and LIPs suggest that their formation has nothing to do with plate tectonics. Instead, LIPs are regarded as the products of plumes of hot (not liquid) rock that push up through the Earth's interior to the base of the crust.[12] Here, where temperatures average 1,200°C, the plume head starts melting and the liquid rock it forms forces its way up through the crust and, wherever it can, extrudes on to the Earth's surface. Where the plume is comparatively small, it may produce only a quite narrow conduit of magma, of a kind that produces lines of islands above a fixed 'hotspot'. But where the plume is larger, the swollen plume head may itself force its way up through the crust and break out at its surface in a great flood of lava, creating a big LIP.

One LIP seems not to fit this bill. This is the Ontong Java Plateau (OJP) in the south-west Pacific Ocean, which has the distinction of being the world's largest LIP (1.5 million km^2, around 0.8 per cent of the entire surface of the planet or the size of Alaska!), which alone perhaps signals an origin distinct from other LIPs. Scientific investigations of the OJP have highlighted several anomalous attributes. One is the distinctive lack of vesicles (small air-filled chambers) within OJP lavas, indicating they erupted beneath at least 800m of ocean water. The problem with this is that, when you apply mantle-plume theory to the origin of the OJP, it should have formed in much shallower water, even hundreds of metres *above* the ocean surface, but its lavas show this just did not happen.

Another odd thing about the OJP is its root. All LIPs have underground roots that represent the conduits, ancient or modern, along which the magma that created the LIP once travelled on its way to the Earth's surface. The OJP has a root that is traceable at least 300km below the ocean floor. According to the various inferential ways we have of examining the composition of such roots, this one should be really hot, perhaps 400°C hotter than the rocks surrounding it. And if it were so hot, then the surface of the OJP itself should today be covered with active volcanoes and fissures along which magma is welling up. But this is not happening. The OJP is clearly an ancient feature, the final days of its activity having been around 120 million years ago.

An ingenious explanation has been proposed for the origin of the OJP.[13] Far from being a true LIP and resulting from processes originating solely within our Earth, it may be the product of the collision with Earth of an extra-terrestrial body (a bolide) maybe 20km in diameter. It does not require a lot of imagination to see what effects such a body moving at perhaps 20km every second might have if it hit the Pacific Ocean today. First, the entire water column would be vaporised. A hole 60km deep would be made in the ocean floor at the point of impact. Instantaneous depressurisation of the rocks surrounding this crater would cause them to melt. Liquid rock would fill the crater and spill out over its edges. The result might resemble a giant LIP.

If all this talk of LIPs and impact craters seems unduly obscure, it is worth perhaps pondering some very human analogies. LIPs are akin to acne in the sense (as many teenagers know well) that they appear from within on the surface of the skin, seemingly without any external stimulus. And like acne spots, it seems LIPs can stay active for an agonising age. But the OJP is different: perhaps a result of a gunshot, followed by bleeding and the formation of a scar that will endure far longer than an adolescent surface blemish.

Land cannot only hide beneath the ocean surface or become an apparently seamless part of other landmasses, it can also burrow beneath these. Rather like a cartoon mouse burrowing under a carpet while the puzzled cat that had been chasing it looks around, wondering where its prey has gone.

We know that in certain parts of the Earth's surface, a chunk (or plate) of oceanic crust is being thrust downwards beneath another chunk of crust of continental origin. Where this happens, any irregularities (like islands) on the down-going crustal plate may find expression in the surface of the continent above long before they reach depths at which they start melting. A good example comes from New Zealand, specifically the easternmost point of North Island, which lies close to the axis of the Hikurangi Trench, where the Pacific crustal plate is being forced beneath that on which North Island lies. Here for some 50km out from the coast, the ocean floor is covered by the Ruatoria landslide complex believed to have formed in the wake of the movement of a seamount – a submerged volcanic island – down the Hikurangi Trench and under the continental crust of North Island. Look at Figure 7.1 and you can see where that seamount, the Horoera Seamount, is thought to be today.

For young June McLauchlan living at the Tatapouri Hotel, some distance inland of the beach on the east coast of North Island, 26 March 1947 was 'a day I'll never forget'. About to walk out of the hotel's front door, June noticed 'the sea lapping at our front lawn' and immediately alerted her parents. The family ran uphill and there watched

> ... first one, then another tsunami race across the ocean and smash onto the land. The first wave took everything other than the hotel out to sea with its backwash. We could see a shed that was full of furniture, a small dinghy, a two-roomed cottage ... then came the second wave, and it dumped everything back almost where it came from. But everything was smashed.[14]

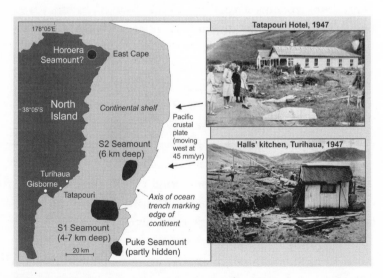

Figure 7.1 *Ancient islands or seamounts being thrust beneath the New Zealand continent cause earthquakes and tsunamis.*

A little further north at Turihaua, a four-roomed house belonging to a Mr and Mrs Hall was also hit by this tsunami. Fortunately, the Halls were in their kitchen – the only room to remain standing after the first wave – with three visitors including a Mr and Mrs Tunnicliffe. The kitchen was 'carried about three metres inland and twisted about by the wave', which appeared to them as a 'large greenish-grey wall' 6m in height. According to an account of their ordeal, which all survived, Mr Hall in the kitchen 'was immersed to the neck and clung to the mantelpiece with one hand while he grasped his floating wife with the other. Mrs Tunnicliffe clung to his back.'[15] Tellingly, no one here felt the earth shake in advance of the tsunamis.

The 1947 tsunamis here were likely caused by an underwater landslip, which in turn was triggered, not by a large-magnitude earthquake of the kind that sometimes occurs along the Hikurangi Trench, but by prolonged low-intensity shaking – a way of gradual rather than sudden energy release.[16] Lasting two weeks, a comparable 'slow-slip' event occurred in this area in September 2014 and involved 15–20cm of movement barely noticeable by people yet detected by a seismic monitoring network. Mapping of the low-magnitude tremors showed they were concentrated around two ancient seamounts (S1 and S2 in Figure 7.1) lying beneath the continental crust of New Zealand, implying that the subsurface under-thrusting of these has a dominant role in causing earthquakes and tsunamis in this area.

Buried seamounts are not only a feature of convergent plate boundaries in New Zealand, but almost everywhere these are found. There are examples along much of the western seaboard of North and Central America where buried seamounts have been identified through their gravitational signals that are generally higher than the sediments burying them. Off the Cascadia margin of North

America, along the coasts of California, Oregon and Washington, the presence of these seamounts is sometimes more difficult to detect because of the huge volumes of river sediment covering them. It is a different story in Central America where rivers are shorter and the continental shelf is narrower. Here imaging of the ocean floor shows seamounts, just like cartoon mice, burrowing under the carpet of continental crust.

Yet when we consider all the causes of land loss within the recent past, it is changes in the ocean surface that explain most. This fact requires some explanation about why and how the sea surface changes, the topic of the next chapter.

CHAPTER EIGHT

Earth's Watery Shroud: Sea Level Changes

In the final encounter between Captain Ahab and the white whale he named Moby Dick, the Captain repeatedly harpoons the whale, but the rope becomes entangled about the Captain's neck and he is carried off into the depths of the Pacific. The fate of the beast and the man entwined in death as it had become in life. The whale also holed Captain Ahab's ship, the *Pequot*, causing it to sink with all its crew. Bobbing in the water, Ishmael, the sole survivor, described the aftermath.

> … small fowls flew screaming over the yet yawning gulf; a sullen white surf beat against its steep sides; then all collapsed, and the great shroud of the sea rolled on as it rolled five thousand years ago.[1]

There is something unsettling about the sea. We often describe it as limitless, as a power we cannot ever hope to subdue. We accept its peregrinations, we bow to its majesty. As we saw in Chapter 4, until quite recently in human history, it may be that drowning people were never saved by others who feared this was to deny the ocean its due. Sometimes drowning people were even encouraged by onlookers to cease their struggles and accept their fate, the sea's life-ending embrace. How outrageous such sentiments seem in today's world … yet they do illustrate people's recently changed relationship with the ocean, a change

underwritten by our hugely improved understanding of
the watery parts of the planet.

We inhabit a drowned world. Science has shown this to
be true beyond doubt. Within the last two million years or
so, even just the last 150,000 years, the average level of the
ocean surface has been some 50–60m below what it is
today (Figure 8.1).

Yet do not imagine for one moment that science was first
to uncover this fact for, as we saw in previous chapters, our
ancestors lived through times when the ocean surface was
lower, at times more than 100m lower, than today. They
survived its rapid post-glacial rise but not without both
discomfort, as shown by the surviving stories of displacement
and involuntary adaptation from the fringes of ice-age
Australia and Kumari Kandam, and inconvenience as
manifested by the evidence of submerged coastal cities,
from 5,000-year-old Pavlopetri to even possibly earlier
ones like Cantre'r Gwaelod and Ys. It seems reasonable to
suppose that the fragments of ancient stories and myths that

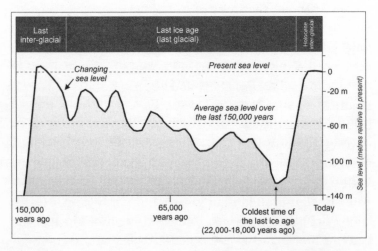

Figure 8.1. *Changes in the ocean surface over the past 150,000 years.*

have reached us today represent memories of times when the ocean surface was much lower than it is today.

We may privilege the comforting rationalism that science provides about past sea level changes, but there is no question that knowledge about these from millennia-old eyewitness accounts was available to us far earlier. So overgrown were these accounts with the vines of narrative embellishment that we failed to recognise them for what they were.[2] But perhaps, just perhaps, in some obscure set of synapses in the rafters of our brains, ancient memories linger about times when the sea level was so much lower. Memories we might one day dust off and activate to help us rationalise the renewed drowning of the world we inhabit.

Over long time periods – centuries, millennia and more – the ocean surface has rarely been static. It is not difficult to understand why. A bucket of water, a swimming pool, a lake, the world's ocean – the principles are the same. If you add more water, the water surface rises. If you throw a few stones in the bucket, 50 people in the swimming pool, a mountain or two in the lake, or extrude a line of volcanoes along the floor of the ocean, the water surface will also rise because the capacity of the container has been reduced. All that pretty much explains why many lands, once occupied by people, have become submerged. Either the ocean had more water added to it … or its water-holding capacity was reduced. Sometimes both things happened together.

Over the past 200 years or so, the sea level has been observed to be rising along many coastlines. The vehemence with which some people deny this to be the case may give you pause for thought, but you should not be fooled. Such people are generally far less interested in what is happening because of the inconvenience this represents, given who they are and what they do. It may be that a person's investment in a cliff-top or beach-back mansion is threatened by sea level rise, but that is not an easy thing to blame, let alone from which to argue for compensation. Far better to deny that sea level is rising and blame the authorities for some supposed oversight that is apparently having the same effect.

More common among deniers of sea level rise is the idea that it cannot be happening because climate change is assuredly a myth, a ruse cooked up by economic aggressors or unprincipled researchers to improperly advantage themselves. I often hear such complaints first hand. But at the end of the day, the cause is less important than the effect, something of which people sometimes lose sight. For there is no question that sea level is rising – a global average of 3–4mm every year currently – and we cannot logically deny that fact because we are sceptical about its possible causes.

If we ignore the short-term oscillations of the ocean surface, like the daily journey it makes between low and high tide, then it has an average level known as 'mean sea level'. During the last great ice age, around 20,000 years ago, averaged across the planet, mean sea level was around 120m lower than it is today. This was because so much water had been sucked out of the oceans to build massive land-based ice sheets that smothered many northern hemisphere continents at this time. Not literally sucked, of course. It is just that in most parts of the world today, when ocean water evaporates and then falls on the land, it does so as rain, which gets into river systems and finds its way back to the ocean where the cycle starts all over again. But imagine if it were so much colder on land that the evaporated water fell there not as rain, but as snow or even ice. That snow and ice could not easily find its way back to the ocean so it accumulated on the land, eventually causing the ocean surface to start dropping. This explains why the ice ages were times of comparatively low sea level and the times between them, commonly known as inter-glacial periods, were times of comparatively high sea level – like that we live in today.

A few years back, before scientists began talking about the Anthropocene, the age in which human influences on the Earth's climate have become dominant, it was the Holocene inter-glacial in which we lived. A time of anomalous warmth and unusually high sea level, which, while not without precedent, was clearly not the average condition of the Earth during the 200,000 years or so of its modern human residency. It is a conceit to think otherwise, unfortunately a conceit that lay unquestioned for many years at the foundation of modern geoscience and its bibliolatrous roots and which informed early understandings of how the recent history of our planet unfolded. The legacy of this belief lives on in the reluctance of science over the past 50 years to accept and then embrace the manifest evidence for global change. Today, there is an overwhelming scientific consensus about the fact of our rapidly changing world, but just a few decades ago some respected scientists were dreaming up convoluted explanations as to why this was not the case.

After the last ice age ended, land-based ice began melting and vast quantities of meltwater poured into the ocean, raising its level. This explains the general rise of post-glacial sea level shown in Figure 8.1 but not all its intricacies. For example, there were short-lived periods of rapid sea level rise representing times when massive lakes of meltwater, unable for millennia to escape the hollows they had ground out for themselves in the centres of continents, finally did so. We saw examples of the effects of this when we discussed

Lake Agassiz in eastern Canada in Chapter 6, but here is
another fascinating example.

The longest river to drain into the Pacific Ocean from
either North or South America is the Columbia, which
meets the sea in north-west Oregon (USA). The first known
person to travel the entire length of the Columbia River
was the mesmeric David Thompson[3] who, upon reaching
Wallula Gap (as we know it today) in its lower quarter on 5
August 1811, described its unusual landscape thus:

> These lands [comprise] pillar like rock, this is often
> like the flutes of an organ at a distance, its strata seems
> perpendicular and is often split in pieces. The pillars are
> split also in various directions as if broken or cracked
> by a violent blow … The surface of the upper rocks
> forms what is called the plains. This is covered with pure
> sand through which the rocks appear everywhere …
> One must say that the finger of the Deity has opened
> by immediate operation the passage of this river through
> such solid materials.[4]

It is astonishing that more than two centuries ago, an
observer without any of our modern knowledge about
Wallula Gap could sense the almost unimaginable violence
of the catastrophic event that formed it. For the time at
which the 'finger of the Deity' poked through these hills
was after the last ice age ended, when a massive ice-dammed
lake to the north-east abruptly broke out of its confines,
sending a wall of water 100m high rushing across the
landscape. Given that this lake – known to geologists as
Lake Missoula – reformed multiple times, there were many
such breakouts and as many as 40 episodes of mega-floods
that drove across this part of the north-west United States.

Not only did the Missoula floods carve out the deep
gorge of the Columbia River, punch through the Wallula

Gap, but they also created one of the most enigmatic landscapes in North America – the Channeled Scablands of eastern Washington state. Here is an evocative description:

> A traveler entering the State of Washington from the East crosses a flat-to-rolling countryside of deep fertile soil commonly sown with wheat. Continuing westward, one abruptly enters a deeply scarred land of bare black rock cut by labyrinthine canyons and channels, plunge pools and rock basins, cascade and cataract ledges, and displaying ragged buttes and cliffs, alcoves, immense gravel bars and giant ripple marks. The traveler has reached the starkly scenic 'Channeled Scabland' and this dramatic change in the landscape may well cause one to wonder 'what happened here?' The answer – the greatest flood [ever] documented.[5]

The story of how a great flood – or rather a succession of great floods – eventually became accepted as the correct explanation for the origin of the Scablands and other features of the Columbia watershed is salutary reading for anyone curious about the weaknesses of scientists and the ways in which scientific revolutions invariably occur.

For much of the nineteenth and twentieth centuries, progress in the geosciences was guided by the doctrine of uniformitarianism, the idea that the kinds of natural processes we see operating in the landscape today are essentially those that operated in the past. Forever. This doctrine, which ironically was developed by dour Scottish geologist James Hutton in 1785 in counterpoint to earlier biblically informed views about Earth-surface evolution through periodic catastrophe, came to dominate the embryonic science of geomorphology to such an extent that it was seen by its most ardent practitioners as excluding any role for catastrophic events in landscape changes.[6]

We live now in more enlightened times, as you might expect, but it was not thus until comparatively recently.

In 1909, J. Harlen Bretz, then a Seattle schoolteacher, started asking questions of the United States Geological Survey about the origin of the Channeled Scablands and, receiving no clear answer, nothing that seemed to adequately acknowledge their uniqueness, he undertook a PhD and wrote a series of scientific articles proposing a catastrophic origin for the Scablands, one which involved a massive flood, 'a debacle which swept the Columbia Plateau'. Like many far-sighted, even courageous, scientists, Bretz stuck to his theory despite a blizzard of criticism from geologists whose faith in uniformitarianism was unflinching. Opposition gradually fell away until at the age of 97, the acceptance of Bretz's ideas was sealed with the award of the Penrose Medal by the Geological Society of America, a triumph that saw Bretz quip to his son that 'all my enemies are dead, so I have no one to gloat over'. Today, while we know that Bretz's insistence on a single Missoula flood instead of many was premature, his legacy as the person to correctly infer – as did David Thompson with his appeal to a divine digit more than a century earlier – a catastrophic origin for the Channeled Scablands and other features of the Columbia River Valley remains unchallenged, a victory that sent ripples through the geosciences, allowing catastrophism to be admitted once more as a significant cause of landscape change.

Each generation of coastal people would surely have noticed what was happening when, between about 15,000 and 6,000 years ago, the ocean surface rose 120m and coastlines in almost every part of the world were progressively – occasionally rapidly – drowned. Maybe, like we see along many coasts today, the places where their grandparents had been born and with which they had been most familiar were all underwater. Land loss may have been lamented, as we saw in Chapter 4, but it also forced people out of coastal zones and inland or offshore to find new places to live. For example, along the Pacific coasts of China and other East Asian countries, it is clear that thousands of years ago post-glacial sea level rise so disrupted coastal living that some people found the resolve, maybe courage, to sail east in search of new lands, something that ultimately led to the first occupations of remote Pacific islands.[7]

Evidence for what seems to have been the very first successful voyage across thousands of kilometres of the western Pacific comes from research by Hsiao-chun Hung and Mike Carson on the islands of the Mariana group in Micronesia. The earliest pottery to be made here was decorated and moulded in ways that are astonishingly similar to how it was being produced at the same time 3,500 years ago in the islands of the Philippines, more than 2,300km of open ocean away.[8]

The likelihood that the Mariana Islands were in fact first settled by people from the Philippines, probably Luzon Island, is also borne out by both genetics and by language. The indigenous peoples of the Marianas – the Chamorros – have DNA lineages that can be traced back to Southeast Asia and appear today as remnant groups in parts of the Philippines; 'a small founding population had reached and settled the Marianas' 3,500 years ago and 'developed unique mutations in isolation'.[9] And linguists have demonstrated that the ancestral Chamorro language in Guam diverged from the Malayo-Austronesian language group within the Philippines, marking these islands as the likeliest homeland for the earliest Pacific Islanders.[10] Pretty compelling evidence that people acquired a permanent foothold in the remote islands of the western Pacific more than three millennia ago.

All that remains is to ponder the implications. That some 3,500 years ago, when people everywhere else on our planet had barely evolved the ability to sail out of sight of land, there was a successful voyage involving a cross-ocean journey of at least 2,300km that brought ancestral Filipinos to a new home in Micronesia. What could have possessed them to make such a journey? How could they possibly have survived three weeks or more at sea, not a jot of land to break their journey? We now have informed answers to such questions.

Imagine a time 4,000 years ago in the Philippines, perhaps in the Peñablanca area of north-east Luzon Island, when most people lived along the coast, sustained by the rich bounty of ocean foods they obtained from coral reefs and lagoons, supplemented by the wild tubers they unearthed from the forest floors, and the fruits and nuts they gathered from the forest trees.[11] They manufactured pottery from local materials, they chewed narcotic betel nut – something that had sustained their ancestors on their sea voyages south from Taiwan – and hunted feral deer and pig. Not to mention the bats that hung in their thousands from the ceilings of the limestone caves that people first occupied here almost 20,000 years earlier. But then a combination of rising sea level and perhaps a succession of extreme waves put an end to this coastal idyll, drowning or washing away places where people had lived for generations and, in doing so, forcing them to move elsewhere. Some moved inland, especially into the fertile yet already quite crowded Cagayan Valley, but others, seeing the difficulties this entailed, were determined to search across the ocean horizon to the east for the islands they felt certain existed there.

The combination of the disappearance of your land and the absence of anywhere else nearby to move is a powerful motivator for contemplating radical change. And without presuming to put ourselves in the minds of people 4,000 years ago in the Philippines, it seems plausible to connect just such a loss of coastal land with a heightened motivation to explore offshore in search of new places to live.[12] There is hardly any land within a few days sail to the east of eastern Luzon so there must have been many unsuccessful attempts to find land, many bamboo rafts crammed with people, which capsized hundreds of kilometres from land. But there was at least one, perhaps several, successful voyages that saw the first human footprints made in the gold sandy beaches of the Mariana Islands. Amazingly, there is evidence by which we can infer that this small success was probably embedded within larger voyaging failure.

This evidence comprises the difference between the things – what archaeologists call material culture – that people had (or soon made) when they reached the Marianas and the things they had back in the Philippines. It seems that the first successful colonisers of the Marianas had no domestic animals, like the dogs and chickens that were a feature of life back home, nor did they have the diversity of pottery forms that were being made and used in the Philippines at the time. This suggests it was a small group, perhaps an offshoot of the main one in the Philippines or one that did not represent the full range of community expertise, which eventually settled the Marianas.

Neither did colonisers have or make bark cloth beaters, used in the Philippines and later in most other Pacific Island groups for the manufacture of cloth from tree bark. Nor, it appears, did the Marianas colonists have the ability to make large pots or ones with lids and handles. All these points favour the idea that the colonising group represented one that had not fully embraced contemporary Filipino culture, perhaps because they had been marginalised, living along its fringes, possibly more recent arrivals who found integration

difficult and eventually abandoned their attempts to take
their chances at sea.

But two things the first colonists of the Mariana Islands
did bring with them were viable seedlings of ironwood,
valued for building and as firewood, and betel nut. The
latter is a fruit commonly chewed in many Pacific societies
today as a stimulant, its bitterness offset by being wrapped
in the betel leaf, dusted with lime powder and rolled with
tobacco. Not only does betel nut consumption give its
devotees crimson mouths, which may startle the unwitting
outsider, but also a narcotic high, which helped sustain
Pacific Island peoples on lengthy cross-ocean voyages
during which less sleep (more alertness) may have equated
with better chances of survival, especially in stormy
unfamiliar landless seas.[13]

A similar story applies to the South Pacific islands, the
earliest occupants of which set out east from bases in the
offshore islands of Papua New Guinea about 3,300 years
ago, eventually settling a succession of island groups from
the Solomon Islands through to Samoa and Tonga within a
thousand years or so. These voyagers not only used betel
nut to stay alert on long ocean voyages but also chewed
kava, the consumption of which has become a defining
feature of many island societies in the region today. As
shown in the colour picture section, there is evidence that
kava was used and prepared by prolonged mastication
among the earliest people in Fiji almost 3,000 years ago,
much as it is today in rural parts of neighbouring Vanuatu.

Imagine you are in eastern Spain, somewhere like Valencia,
and you are enticed into a smart seafood restaurant to lunch
on their special mussels-and-seafood spaghetti. Or you are
a little further south, perched on a high stool propping up

the bar at a beachside *chiringuito* and, feeling peckish, order a plate of tapas that includes steamed clams, freshly harvested. You are actually not so different to the people of this area 9,500 years earlier who also consumed shellfish from the ocean shallows along its coast. We know that the protein intake of these Mesolithic Spaniards comprised at least 25 per cent marine foods, some way off dependence yet still significant.[14]

In this part of Spain, about halfway between Benidorm and Valencia, lies the Pego-Oliva Marsh around which people, both Mesolithic and later Neolithic, lived for 10,000 years or more.[15] Around the time they arrived, the sea level here was more than 20m lower than today so, in order to understand how people lived here that long ago, researchers have cored the soft sediments that comprise the coastal plain, onshore and offshore, to reveal evidence of past environments. For much of the post-glacial period, up until 7,000 years ago, the sea surface here rose at an average rate of 9cm each year. This equates to an average shoreline recession of 360m every 100 years for at least three millennia. A Mesolithic woman who lived to be 50 would have witnessed almost two football fields of coastal land loss in her lifetime. What must people have thought was happening?

We infer that the Mesolithic peoples of Pego-Oliva eventually became so frustrated, so disturbed, by these rapid coastal changes, particularly the ways they made finding food ever more difficult, that around 8,000 years ago they left the area. Following a 700-year hiatus, it was repopulated by Neolithic farmers who arrived, perhaps by design, at the time when the pace of sea level rise in the western Mediterranean was slowing and would soon stabilise. Planting wheat and barley, they flourished. Their descendants may still live in Pego-Oliva.

The picture from eastern Spain of rising sea level over several millennia causing a rapid landwards movement of the shoreline contrasts with that from the Yangtze River mouth in eastern China. Here the effects of rising post-glacial sea level in drowning the fringes of the land were countered by the river water and sediment carried downstream and dumped along the shoreline. In response, the shoreline here has shifted back and forth for much of the past 7,500 years in a swaying dance that alternately delighted and frustrated the audience. Sometimes shoreline changes created new opportunities for people, leading to prosperity and a burgeoning of culture, then took these away, spawning deprivation and cultural decline. Key to these swings was lowland rice cultivation, an undisputed driver of cultural complexity, which led to the building of the earliest towns and cities in this part of the world.

A hundred kilometres or so south of the Yangtze River mouth, past the sprawling Shanghai metropolis, across Hangzhou Bay (spanned by an almost unreal 36km-long bridge), one reaches the small agricultural settlement of Hemudu. Where civilisation, as we define it in much of the modern world, may have first stuttered its way to an enduring start. The modern story began in the summer of 1973 when Hemudu farmers were excavating a pit for a water pump and, finding pottery fragments and animal bones, alerted local authorities. Excavations subsequently demonstrated that rice was being grown at Hemudu as much as 7,000 years ago but, most importantly, clear signs were found that the earliest inhabitants of Hemudu lived in an organised settlement comprising numerous buildings with different functions surrounded by a wooden palisade. Hemudu represents the earliest-known complex settlement, the development of which is thought to have been possible only because of rice cultivation.

Although in Chapter 6 the possibility that rice was domesticated in now-undersea Sundaland was discussed, rice is widely regarded as having first been domesticated in the lower part of the Yangtze Valley around 10,000 years ago. It provided a solid platform for sustaining population growth and eventually increasing societal complexity. Some of the earliest cities anywhere in the world – places like walled Liangzhu that date from 3,300 years ago – were built in the Lower Yangtze, a surplus of rice allowing food to be supplied to non-producers, people like priests and warriors, monarchs and merchants. The start of a social compact in which many of us today are part, consuming food grown for us by others.

Along most of the world's coasts, the post-glacial rise of sea level ended 6,000 years ago, but in parts of north-west Europe it continues still. The reason is because these areas continue to be affected by changes in land level induced by the removal of nearby heavy ice loads after the ice age ended. As shown in Figure 8.2A, when ice sheets develop on a continent, the underlying crust is pushed downwards and slowly flows into peripheral areas causing these – like a tightening belt around a substantial waist – to bulge. After the ice melts, as shown in Figure 8.2B, it is as though the belt is being loosened. The crust over which the ice once lay rises while the bulges gradually collapse.

This process, known as isostatic readjustment, explains why the land is rising today in formerly ice-covered areas like Scotland, Scandinavia and much of the Canadian Shield. So fast is it rising that in some places it outstripped the rate of post-glacial sea level rise such that ancient shorelines, which elsewhere in the world lie beneath tens of metres of ocean, are found on dry land.

An example comes from Haida Gwaii (the Queen Charlotte Islands) off the west coast of Canada. Among indigenous Haida stories is one about a great inundation,

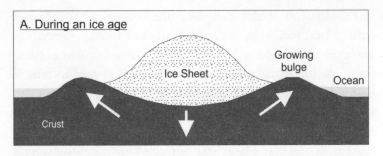

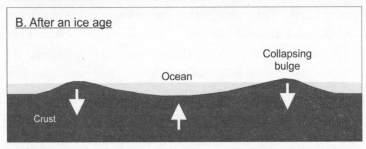

Figure 8.2 *A – During an ice age, ice sheets that form on land press down on the Earth's crust below, causing it to flow outwards to form peripheral bulges. B – After ice sheets melt, the weight on the crust is relieved and it rebounds upwards while bulges collapse.*

mentioned in Chapter 1, which may recall events that took place 9,700–12,700 years ago. Coastlines from that period remain under the ocean, but 9,700 years ago, because of isostatic readjustment, the land here started rising, carrying younger coastlines – and information about the people who occupied them – upwards to places that are now above present sea level. For example, within the rain-misted forests that cloak southern Haida Gwaii are ancient shorelines, some 15m above sea level, dating from 3,000–9,200 years ago. At the Arrow Creek site, occupied 5,600–8,100 years ago, hundreds of stone tools from this time have been recovered. They show that the coastal Haida of the time not only processed and ate shellfish, caught fish and hunted large-bodied prey such as bear and caribou, but also felled cedar and spruce trees to make frames for hide boats that they used to journey back and forth along the coast. Haida Gwaii is one of the very few places in the world where we can easily access this kind of information from this time period. Almost everywhere else, coasts this old are underwater.

As formerly ice-depressed areas rose after the end of the last ice age, the peripheral bulges also began collapsing, the surface of the crust levelling out once again, as shown in Figure 8.2B. This means that in places adjoining those once covered by ice sheets, the land is today sinking. This is what is happening along former ice-sheet margins, be they in Chesapeake Bay on the US east coast (discussed at the beginning of Chapter 3) or in those parts of north-west Europe where sea level has continued to rise since the end of the last great ice age. It is perhaps no surprise that the oldest cultures of north-west Europe are unusually replete with ancient stories about recently submerged lands and 'sunken cities', as noted in Chapter 4. In fact, the continuous submergence of these coasts for 15,000 years may even account for the 'brooding pessimism' said to characterise some of its modern cultures[16] and explain why others in this part of the world apparently learn to be miserable.[17]

Far-fetched, of course, but consider how it must have been for aspirant tyrants occupying the Atlantic fringes of north-west Europe within the past few thousand years. Every generation, every century, every dynasty mobilised the power and the resources needed to build something big and intimidating and memorable along the coast. But then, unlike the situation in Germany or Italy, or even further afield, this enterprise was struck down – heartlessly, it seemed – by the rising ocean. They didn't know it at the time, but this spelt the end of their era of influence. Within a few hundred years, memories of their life's work would be as forgotten as the coastal cities they constructed. The ocean was no friend to ancient coastal enterprise in north-west Europe, which may be why many of its oldest capital cities – like Berlin and Brussels, London, Madrid and Paris – are inland and not on the coast.[18]

The human implications of the contrast between coasts of north-west Europe, where sea level has been rising continuously for 15,000 years, and most coasts elsewhere in the world, where post-glacial sea level rise ceased about 6,000 years ago, is worth delving into some more.

Let us start with an example from the latter, globally more common, type of situation. Good examples come from Australia, one of the few fragments of continental crust that today does not have any plate boundary along its fringes. Australia is majestic, aloof, having been carried north-east on its eponymous crustal plate for some 30 million years, memories of a wild misspent youth buried in ancient rocks.

During the last ice age, the ocean surface far lower, Australia was connected by dry land to what we now call Papua New Guinea (a situation shown in Figure 6.3). People would have occupied this low fertile region in great numbers, but after the ice age ended and sea level started rising they were gradually pushed out into the higher lands of New Guinea and north-east Australia. It may be no coincidence that the most widespread group of indigenous languages in Australia, the Pama–Nyungan, once spoken across 90 per cent of the continent, appears to have originated somewhere in north-east Australia. Quite conceivably, the roots of this language group lie in those spoken by the people who occupied the ice age land bridge between Australia and Papua New Guinea before being dispersed by rising sea level.[19]

Up until then, life on the land bridge had been quite good. Lacking many high points – those that existed are now the scattered islands of the Torres Strait – the land would have been readily traversable and presented consistent opportunities for subsistence. People would have roamed its savannas, gathering nuts and fruits, hunting the animals with which they shared the land. But life centred around the vast freshwater lake – Lake Carpentaria we call 'it in retrospect – that never dried out. Connected to a series of freshwater swamps, home to numerous species of bird, the lake was a magnet for people.

Yet, as life in this antipodean Elysium showed signs of decline, every year the shoreline moving landwards creating coastal squeeze, so its occupants – human and other – began moving out. Around 11,700 years ago, the freshwater Lake Carpentaria became connected to the rising ocean – an event with rapid and dramatic ecological consequences. Within a few hundred years, the memory of the lake must have faded as it morphed into the Gulf of Carpentaria, which dominates the coastline of north-east Australia today. As the sea rose along the shores of the gulf, so they became inundated. What were once mainland promontories had their necks severed by the sea, creating islands where none existed before.[20] How this affected people and their activities is largely uncertain although a number of Aboriginal tribes living in this area have, remarkably, retained memories of these events for more than 7,000 years.

Dick Roughsey, who got his English surname from his Lardil (Aboriginal) name Goobalathaldin, which means 'rough seas', was a renowned artist, born in 1920 on Mornington (or Gununa) Island, part of the Wellesley group, in the southern Gulf of Carpentaria. Largely through his lively autobiography, he is also remembered as the person who brought Lardil stories to the world's attention. One of these recalls when Mornington (and the other islands in the Wellesley group) were joined to the Australian mainland.

In the beginning, our home islands, now called the North Wellesleys were not islands at all, but part of a peninsula running out from the mainland. Geologists ... thought

that the peninsula might have been divided into islands
by a big flood which took place about 12,000 years ago.
But our people say that the channels were caused by
Garnguur, a sea-gull woman who dragged a big walpa or
raft, back and forth across the peninsula.[21]

The most astonishing thing about this story is that it is
likely to have endured for at least 7,450 years, the most
recent time at which the islands could have been joined to
the mainland, passed on orally across generations of the
Lardil until it was first written down 50 years or so ago.[22]
But there is a detail in the story – the dragging of the large
raft back and forth across the neck of the peninsula – that
describes so exactly the (now invisible) corrugated form of
the sea floor in this area that one can barely doubt the story
was based on close and sustained observation of the slow
submergence of the neck of the peninsula seven millennia
and more ago.

Lastly in this chapter, we turn to the anomalies, the
coasts along which sea level has risen unceasingly since the
last ice age ended because they are located in areas where
peripheral bulges (shown in Figure 8.2) continue to
collapse. And where better to pick an example than along
the Atlantic coasts of north-west Europe where stories
about submerged cities, such as Ys (discussed in Chapter 2)
and Cantre'r Gwaelod (in Chapter 4), abound? Another
example is the land of Lyonesse, a place with Arthurian
connections said to have been located off the western tip
of Cornwall (UK) – in most versions including the
Scilly Isles.

A key question is whether the Lyonesse tradition has, in
the 1927 words of archaeologist Osbert Crawford, 'any real
basis in fact, or is it merely an invention of the dreamy
Celt?'.[23] Given that few scientists have seen any merit in
stories about Lyonesse, it is hardly surprising that the

overwhelming view is that they are fictions, a bunch of fantasies put together by dreamy Celts. This strikes me as implausible, a shallow sentimental default judgement. If 'dreamy Celts' wanted to invent stories, why not those exaggerating their achievements, asserting their invincibility, rather than a somewhat inward-looking doleful story of land loss? Unless of course, land had actually been lost and the story was the only medium available to the eyewitnesses through which to communicate it to future generations.

Some versions of the story say that Lyonesse lay *between* Cornwall and the Scilly Isles, a distance of some 40km in which ocean depths reach 65m, but Crawford favoured the idea that Lyonesse was actually the Scilly archipelago at the time its larger islands were all joined (Figure 8.3). The idea of Lyonesse as having been the former, much larger, area might be supported by accounts claiming it contained 140 parish churches and that the sole known survivor was one Trevillian, who escaped on a white horse and later founded an ancient Cornish family, which features a white horse on its coat of arms. Yet the idea that Lyonesse was the Scilly Isles is consistent with Scillonian traditions and observations, made by Crawford and others, of submerged stone walls (known as hedges) and causeways. It seems more plausible.

The last time all the major islands in the Scilly group were joined was at least 7,700 years ago, a time when it had no permanent population. But during the Neolithic Revolution, some 5,500 years ago it had many inhabitants, identified largely through the remains of their distinctive pottery. At this time, the three main islands – St Martin's, St Mary's and Tresco – formed a single landmass, showing that the Scilly-as-Lyonesse stories could indeed be based on memories of its subsequent drowning.[24]

Many stories about Lyonesse link its disappearance to large waves and floods, but these alone, as we shall see in Chapter 11, cannot cause the permanent submergence of

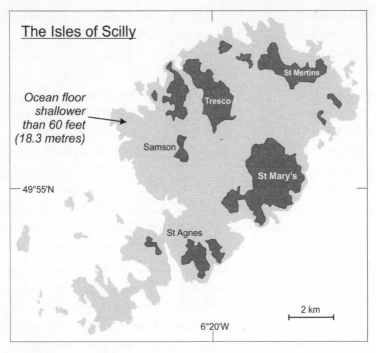

Figure 8.3 *During the time that people have occupied the Scilly Isles, off the south-west extremity of mainland England, its principal islands were joined. Then the rising ocean drowned their shallower parts and created the archipelago seen today. Stories about the sunken land of Lyonesse may have their origin in these events.*

solid land. Yet major floods are often memorable events, considered worthy of recall in pre-literate societies, so it may have seemed logical to link them to older stories about land submergence. One flood in Cornwall, on 11 November 1099, noted that 'the sea overflowed the shore, destroying towns and drowning many persons and innumerable oxen and sheep'. Other accounts explain traditions such as that from 1586 where at Land's End in Cornwall

... the Inhabitants doe Suppose that this Promontorie heretofore ran further into the Sea, and by the rubbish [meaning flotsam] which is drawne out from thence, the

> Mariners affirm the same … that the earth now covered
> there all over with the in-breaking of the Sea was called
> Lionesse.[25]

It is no surprise that survivors of terrible floods sought to associate them with the same processes adjudged responsible for the ancient submergence of Lyonesse. And over time, after perhaps several generations of telling the same story, the events became fused. Again and again in the memories of local residents, ancient Lyonesse was miraculously reborn only to be submerged anew.

The next chapter probes further into the question of land-level changes, and how these have caused places to both rise from the shadowy depths of the ocean and to sink within them.

'The Island Tilts … the Tourists Go Mad': the Ups and Downs of the Earth's Crust

A hundred years ago, off the coast of Sicily in a volcanically active part of the Mediterranean Sea, a party of upper-class English tourists – boaters on their heads, parasols unfurled against the unfamiliar heat – was ferried by local boatmen to land and take tea on an uninhabited island, which they were informed was of 'volcanic' origin.

> 'Pooh, volcanic!' says the leading tourist, and the ladies say 'how interesting'. The island begins to rock, and so do the minds of its visitors. They start and quarrel and jabber. Fingers burst up through the sand – fingers of sea devils. The island tilts. The tourists go mad'.

This fictitious account by E. M. Forster, from his 1907 novel *The Longest Journey*, was intended to demonstrate the philistinism of the English upper classes, and was based on real places and real events in the Mediterranean, as we shall see in Chapter 11.[1]

The land we occupy does sometimes rise, fall, even tilt. Mostly these tectonic movements are so slow they are barely detectable except by the most sensitive instruments, but occasionally they occur more rapidly. Within a few minutes, even seconds. Movements that can radically reconfigure the geography of a place, tempting both onlookers and more distant observers to attribute causation that leans more towards vengeful gods than evidence-based science, traumatising the experience more than necessary.

By the late seventeenth century, the city of Port Royal on the island of Jamaica was thriving. It was the largest city anywhere in the Americas at the time. Fortuitously located, it had become the trade hub of the Caribbean, its prosperity resting on slaves and sugar … and the ill-gotten gains of some of the most notorious pirates in history. Port Royal had gained a reputation as an unsavoury place, 'one of the Ludest in the Christian World, a sink of all Filthiness' in which one-fifth of all buildings were 'brothels, gaming houses, taverns and grog shops'.[2] Most of Port Royal was submerged within two minutes on 7 June 1692 (Figure 9.1).

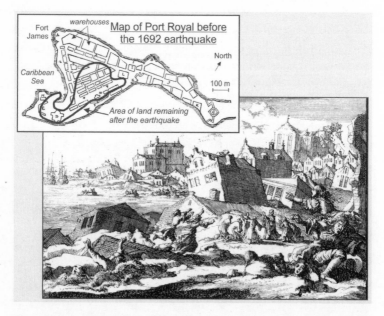

Figure 9.1 *The abrupt submergence of Port Royal (Jamaica) in 1692, painted by Jan Luyken and Pieter van der Aa. Inset shows the extent of land lost.*

This was not a tectonic subsidence – one caused directly by crustal movement – but a result of earthquake-induced liquefaction of the sand spit on which two-thirds of the city had been built. The spit was shaken repeatedly, making the water pressure between the sand particles increase quickly, eventually allowing them to move relative to one another. The solid ground became fluid. As on the fictitious Mediterranean volcanic island described above, this also caused fountaining of water and sand – the sea devils' grasping fingers. The Reverend Dr Emmanuel Heath survived the 1692 Port Royal earthquake and left a graphic account of some effects of the liquefaction. Some unfortunate people, he noted, 'were swallowed up to the Neck, and then the Earth shut upon them; and squeezed them to death; and in that manner several are left buried with their Heads above ground'.

Like others who witnessed the carnage (and many who did not), Reverend Heath concluded of the inhabitants of Port Royal that 'by this terrible Judgment, God will make them reform their lives, for there was not a more ungodly People on the Face of the Earth'.[3] The 1692 submergence of the city and its terrible human toll had a significant impact on the understanding of natural disasters in many parts of the world. It occurred at a time, at the dawn of the Enlightenment,[4] when the degree of divine influence on Earth-surface evolution was being questioned and mechanistic explanations were becoming favoured to explain a whole range of natural phenomena. The Port Royal earthquake provided powerful ammunition for critics of Enlightenment thought who argued that it showed (yet again) the ways in which God controlled the world and punished those who failed to abide by divine tenets. To many of us today, the idea of attributing natural disasters to godly retribution may have a distinctive medieval flavour, yet contemporary examples abound.[5]

Since earthquakes are usually short-lived and their manifold effects occasionally catastrophic, it is often difficult to unpack the precise sequence of events they involved, especially in their immediate aftermath. With hindsight, it is likely that the 1692 Port Royal earthquake triggered an underwater landslide, which abruptly displaced such a large volume of sea water that it caused a tsunami, a giant wave (maybe waves), to wash across the land. Liquefaction, which was a conspicuous feature of this earthquake, rarely occurs after a single jolt, being more commonly the result of a short-spaced succession of earthquakes. It is easy to imagine the main drawing in Figure 9.1 represents the effects of a large tsunami wave washing ashore, but actually it probably better captures the city of Port Royal being washed out to sea as a result of the land on which it was built suddenly turning to slurry.

Just 50km south-east of Anchorage, the largest city in Alaska, along the narrow sinuous inlet known as Turnagain Arm lie the remains of the town of Portage. The place names reveal their original function. The name of the city at the head of the Cook Inlet, sheltered from turbulent Arctic seas, is where ships could safely anchor, one of which was a floating hardware and clothing store named The Anchorage. The aquatic cul-de-sac was named River Turnagain by Captain Cook in 1778 because his ships had to do this when it proved not to be the sought-after Northwest Passage. And at its head lay the place where canoes might be carried (or portaged) across the lowlands into the rivers and lakes that connected Turnagain Arm to Prince William Sound.

On 27 March 1964, in Portage, Alaska, the grey dawn had broken into an unseasonably cold spring day and lots of snow still lay on the ground. The first inkling that 11-year-old J. R. Shackelford had that something was not quite right was at about 5.30 p.m. when his family's 'dog and cat were going crazy at the front door'. He let them out 'and they both took off like bullets. They knew something was about to happen and they didn't want to be in that house'.[6]

The largest-magnitude earthquake ever measured in North America occurred a few minutes later. It shook up Anchorage, literally, and destroyed Portage by dropping much of it abruptly 3m into the icy waters of Turnagain Arm. But the greatest effects of this earthquake occurred closer to its epicentre in Prince William Sound, after which it was named. Lashed by giant tsunami waves, several coastal towns along the shores of the sound were rendered uninhabitable. Parts of Kodiak Island rose by as much as 11m, while parts of Montague Island dropped 2m – all within a few minutes.

Among seismologists, the 1964 Prince William Sound earthquake is famous not only for its almost unprecedented magnitude, at least in recent human history, but also because it occurred as a result of a single low-angle thrust along what we now recognise as the sub-surface boundary between two crustal plates. The origin of this massive quake was first deduced by George Plafker, one of the first geologists on the scene, but his explanations were initially rejected. It was not simply a matter of correct causation, but the global implications of this. For in deducing that a single shallow thrust caused this earthquake, Plafker was implicitly endorsing then-radical ideas about Earth-surface mobility that were challenging more long-standing fixist notions about the evolution of our planet's surface. Plafker stuck with his first-hand deductions and finally, years later, won the day.[7]

This chapter focuses on places, like Portage and Montague Island, which were rapidly submerged by downward land-level movements. These occur typically along convergent plate boundaries where one crustal chunk (or plate) is being pushed beneath another. Earthquakes typically occur in such places because of the release of stress. As one chunk pushes against the other, nothing moves at first because friction prevents it; in other words, the two crustal plates 'stick' together. You can do this with your fists by placing one on top of the other, pressing down and sideways just enough to have them stick together. Increase the sideways pressure just enough to overcome the friction and your fists slip past one another. When this happens along convergent plate boundaries, stress is released and an earthquake occurs; one plate slips a short distance past the other and then sticks again. Whenever there is a slip, the energy released sometimes also causes the Earth's surface above the earthquake epicentre to move either up or down. Sometimes simultaneously, as in the 1964 Prince William Sound earthquake.

Unsolved puzzles about our past that involve the ups and downs of the solid earth are more common that you might expect. Places that once existed appear literally to be lost, no one certain where their remains are today, or indeed what caused their disappearance.

Dating from the third or fourth century BC, the coastal ports of Kerala in western India were so invested in international trade that this province has jokingly been described as having a Western rather than a South Asian heritage.[8] Some of those ports – like Calicut (now Kozhikode) and mysterious Muziris – have disappeared, sometimes with the sizeable towns their activities supported, and not to mention the conspicuous wealth amassed by their rulers. What actually happened remains uncertain so here are the essentials of the stories.

One of the most reliable of the relevant ancient texts, perhaps because it was written by a chronicler unencumbered by an excess of education, is the *Periplus of the Erythraean Sea*, the logbook of a trading voyage about AD 60 that included the west coast of India.[9] The author visited five coastal market towns before arriving at Muziris, some 4km up the Periyar River from the coast, a place which 'abounds in ships sent there with cargoes from Arabia, and by the Greeks'. Muziris had been known hundreds of years earlier and its role in trade was extolled by both Roman and Tamil writers. For example, a verse in the *Agananuru* describes Muziris as the place 'where the beautiful ships of Yavanas [foreigners] bringing gold come, splashing the white foam of Periyar ... and return laden with pepper'.[10] The Roman writer Pliny adjudged Muziris the 'emporium of India', a reference to its role as a supplier of unique exports including 'pepper coming from Kottanara, a great quantity of fine pearls, ivory silk clothes ... transparent stones of all kinds, diamonds and sapphires and tortoise shells'. A map made during the fifth century AD shows Muziris, its Roman temple and military barracks which housed as many as 1,200 Roman soldiers to secure the lucrative trade and provide protection against pirates.

The mystery of Muziris is that its presence was routinely documented up until AD 1341 after which no one mentions having been there. In particular, it failed to be mentioned by the prodigious travellers Ibn Battuta and Marco Polo, who one would expect to have encountered it during their travels through the region. So where was Muziris located and what happened to it?

For a long time it was thought that ancient Muziris became the port of Kodungallur, which was reportedly destroyed when the Periyar River flooded catastrophically about AD 1341 but, since archaeological excavations there failed to unearth any trace of an ancient city, attention switched to the area around the small village of Pattanam, a short distance away. For years, during the monsoon season, children here had entertained themselves collecting colourful beads that rose to the ground surface, a sure sign of buried riches. And in Pattanam indeed, archaeologists found evidence for an ancient port city that had enjoyed a millennium of active trade involving, for the first known time in India, Roman amphorae. But not everyone is convinced that this is Muziris ... not least because of its contemporary rural character, considered an unlikely disguise for an ancient city of such importance: 'tall arcing palms, squat plantains, vines and creepers, near-fluorescent monsoon grass. There are sporadic houses, a temple, a village office and sudden channels of water,' wrote a sceptical journalist.[11]

I have an explanation. In AD 1341, a huge flood of reportedly unprecedented magnitude affected this part of the Kerala coast. Some say it altered the lower course of the Periyar River such that Muziris suddenly found itself hundreds of metres from a navigable river channel, forcing the relocation of its prime functions (and merchants) to Kochi (Cochin), a city 10km south where the same floods had fortuitously opened up sea access where none existed before (Figure 9.2). From the huge amounts of sediment brought down the Periyar by his flood, the offshore island of Vypin at the mouth of this river is said to have been created. Today this elongate island is 25km long, improbably large to have been formed during a single flood, which suggests it owes its origin to something else. And the question should also be asked why, if it is merely a pile of river-deposited sand and gravel, is Vypin Island still there? Why has the combined force of river water and sea waves not erased it from the geography of coastal Kerala, as you might expect?

One answer is that Vypin Island was created not by a flood but by land uplift, as is indeed explicitly stated in

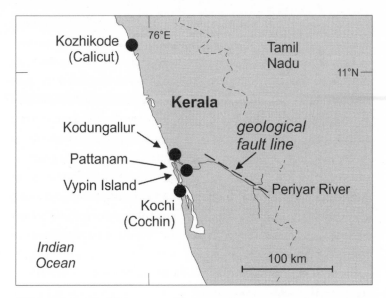

Figure 9.2 *Possible locations of Muziris, the emporium of ancient India.*

some of the earliest written accounts of the area. One, written in 1900, links the formation of Vypin to a severe earthquake, and I surmise this may have echoed the prevailing wisdom at that time.[12] For it is demonstrably possible, as happened during the 1964 Prince William Sound earthquake, for land not only to rise up during such events, but also simultaneously to sink. So, if we allow that Vypin rose up abruptly during the AD 1341 earthquake, maybe it was also the case that Muziris, wherever it was, suddenly sank, erasing it in a short moment from history.

In support of this suggestion to explain the fate of Muziris, let me discuss another example from the Kerala coast. This is the trading port city of Calicut (Kozhikode) that may have been founded in the fourteenth century. When the Portuguese commander Vasco da Gama landed at Calicut on 20 May 1498 to negotiate a trading partnership, he found the city to be inhabited by Christians, something that redoubled his will to succeed. Christianity may have reached Kerala as early as the first century AD, popular traditions recalling it was introduced by Saint Thomas, one of the 12 apostles of Jesus. Whether or not this is true, it cemented the foundations for trade links between Kerala and outward-looking European powers in medieval times. Yet the prospects did not appear too good after Da Gama's visit, for the Zamorin, who ruled Calicut, was unimpressed by the quality of the gifts sent by the Portuguese monarch and rejected Da Gama's proposal to establish a Portuguese presence there. This happened eventually although the later Calicut appears to have become an impoverished descendant of that visited by Da Gama. An earthquake may have contributed to this.

Here is a description of Calicut in 1703 by British sea captain Alexander Hamilton.

In Anno [year] 1703 about the Middle of February, I called at Calecut [Calicut] on my Way to Surat, and standing into the [sea] Road, I chanced to strike on some of the Ruins of the sunken Town built by the Portugueze in former Times. Whether that Town was swallowed up by an Earthquake, as some affirm, or whether it was undermined by the Sea, I will not determine.[13]

The earthquake explanation was echoed by dragoon officer Stanhope in 1785 as representing local residents' memories.

Callicutt, which is now an inconsiderable village, was formerly a magnificent city, and the residence of a powerful Prince. According to the tradition ... it was overwhelmed near 200 years ago by a sudden rising of the sea, and all its inhabitants perished. We anchored on the spot where the ancient city stood and as we went ashore at low water, the foundations of the buildings were discernible to the naked eye.[14]

Both Hamilton and Stanhope infer that this earthquake occurred in the 1580s.

Earthquakes, especially the larger land-surface altering varieties, do not occur everywhere, so it is reasonable to question whether or not the Kerala coast is vulnerable to these. Classified by geologists as part of a stable continental region, you might suppose not. But that would be a hasty judgement because even such places are invariably criss-crossed by networks of geological faults along which crustal stress can build up and then occasionally be released as earthquakes. And some of these can be alarmingly strong and damaging.

In Kerala, as shown in Figure 9.2, part of the Periyar River valley follows a prominent geological fault line that has been the source of some recent earthquakes. This demonstrates the propensity for earthquakes in this area but falls short of proving that coastal submergence occurred during these events. Yet the fact that 'the sources in central midland Kerala have been generating notable earthquakes many times in the past' should at least allow us to consider that apparent abrupt changes in medieval land level at Calicut and Muziris were caused by large earthquakes.[15] Our openness to such explanations has implications for preparing vulnerable populations through, for example, the understanding of the causes and precursors of earthquakes and tsunamis that might one day affect them. We brush the lessons of history aside at our peril.

Despite there inevitably (actually, unavoidably) being some remaining uncertainty, it seems to me most likely that the medieval Keralan port cities of Calicut and Muziris were not washed from existence by exceptionally large river floods, but rather by earthquakes that abruptly displaced fault-bounded blocks of land, maybe liquefacting the loose materials on their surface, while also generating tsunami waves that washed across the land. Memories of these events by those who witnessed them and lived to tell the tale recall both land movements and flooding, but inevitably these memories blur and fuse as time passes. People's instincts and prejudices often determine which explanation subsequently dominates.

Kerala is not the only place in the world where memories of ancient cataclysmic events that had multiple expressions and effects have come down to us today in a form that makes them challenging to understand. In the south-west Pacific Ocean, straddling the boundary between two sizeable crustal plates, lie several island groups in which large-magnitude earthquakes often occur, sometimes producing both abrupt vertical movements of the land surface as well as tsunami waves. There are even some well-attested examples of some entire islands having vanished during such events.[16]

Better known for arresting the notorious blackbirder Bully Hayes and then being bamboozled into releasing him to continue his lawless activities in the Pacific Islands, Commander Richard Meade of the USS *Narragansett* also sailed through the islands of Vanuatu in the late nineteenth century. One day in 1872, the corvette came across some excited local people who told the Commander an extraordinary tale, one he duly recorded in his *Remark Book*. The story concerned an island that had only recently disappeared, the inhabitants of which, wrote Meade, 'saved themselves with great difficulty'. Later research confirmed these spare details, finding that the island was named Mamata and located in the area between the modern islands of Ambae, Maewo and Pentecost.[17]

Mamata rose from the undersea flanks of massive active volcanic Ambae, which one day in 1870 experienced a summit eruption. So violent was this event that it shook the island to its roots, causing a landslide of likely epic proportions on its south-east flank from the underwater part of which Mamata rose. This island was probably submerged in a matter of minutes, the handful of survivors said to have paddled to Baitora on the west coast of Maewo where their descendants still live today. We can imagine the eagerness of the survivors, seeing the *Narragansett* sailing past Baitora just two years later, to relay their extraordinary story to the world beyond.

But it is to another group of islands in Vanuatu that I want to turn to illustrate the ways in which catastrophic land submergence can be unpacked by careful research, a process which sometimes forces traditional explanations preserved in ancient stories to be questioned. These particular stories had been known for ages, kept alive orally until they were first written down in the early nineteenth century.

It was John Layard, Cambridge University scholar, who in 1914 arrived on the islands of Vanuatu (then named the New Hebrides) to investigate rumours of 'a living megalithic culture' on the small islands off the east coast of large Malekula Island. Layard achieved his goal, producing an 816-page book named *Stone Men of Malekula* that compiled details of the Maki, a religious belief system centred around megaliths. Tangential to his principal interest was the recording of stories about vanished islands, the most detailed of which he received from a 'first-class' informant named Mataru living on the tiny offshore island of Vao. At the time of Layard's visit, Mataru was a man 'in the prime of life, who had once been a member of a Presbyterian Mission school [on the main island of Malekula], but who had renounced Christianity and returned to Vao in order to rebuild the fortunes of his family by an intensive prosecution of megalithic ritual'. Mataru was descended from a lineage that originated on the island of Tolamp, which reportedly sank beneath the ocean surface several generations earlier. Some modern stories about sunken Tolamp are related in Chapter 3.

Tolamp is likely to have been a volcanic island similar in size and form to Vao. It is said to have had a flourishing population that played a prominent role in Maki ceremonies before its disappearance, a few hundred years before Layard's visit. A nearby island named Malveveng also disappeared, probably around the same time as Tolamp. Stories of why and how these islands disappeared were told to Layard by Mataru and are still well known in the area today.

Several stories explain that Tolamp was washed away by giant waves whipped up by a man on the Malekula mainland whose daughter was unhappy in her marriage to a man living on Tolamp. Another story recalls that the day Tolamp disappeared, most of its inhabitants had travelled to the mainland.

> One man told his two small boys to stay at home, that they might play close to their house, but must on no account go to the other Side of the Island. After their parents had gone, the two brothers … went over to the other Side. After a while, they sat down and started hollowing out a hole in the sand. Their hands came up against two stones. They uncovered them; they lifted them up. Then suddenly, water began to spring forth, to rise, and finally to gush out of the hole. It was the sea swallowing up the island. The few men who had stayed home immediately rushed and broke off branches in a vain attempt to stop up the hole. But it was useless; the hole grew as they watched. Then there was a panic. All the men of Tolamp who had come back from the mainland gathered together their pigs, their bows, their tools for hollowing out canoes, everything they could, then sailed back over the strait and established themselves on the mainland. Some who had not hurried enough were swallowed up by the flood; others tried to swim and were eaten by sharks. The island foundered beneath six feet of water at low tide.[18]

An entertaining yarn you might say, but it is also one with clues as to what probably caused Tolamp to disappear because it cannot simply have been large waves. Volcanic islands cannot be washed away, but an eyewitness might easily get that impression, especially if they did not feel the undersea earthquake that shook the island and triggered the sea floor landslide that actually made it slip beneath the ocean surface. You might have got that impression if all you witnessed were the tsunami waves, produced by the earthquake, that smashed into the sinking island making it look − for all intents and purposes − as though they were actually destroying it.

There are other examples of islands that vanished the same way and there are commonalities in the stories of the people whose ancestors witnessed these events. There may be similarities between what happened at Tolamp and what happened in the 1692 Port Royal earthquake when liquefaction rather than large waves likely caused the land to disappear. Did the people of Tolamp think the water that sprang forth from the hole the two boys made might have been the fingers of sea devils, perhaps clawing the land back into the ocean depths where it was considered to rightfully belong? And do such interpretations explain why some people, until comparatively recently, believed it wrong to attempt to rescue drowning people, as discussed in Chapter 4, because the ocean was beyond human jurisdiction? It is plausible.

So why and how do islands like Mamata and Tolamp actually disappear? It must first be appreciated that almost every island in the middle of an ocean is actually just the small exposed tip of a massive volcanic mountain rising often thousands of metres from the deep ocean floor where it was born. Sometimes when that mountain is shaken by an earthquake, part of one of its steeper flanks might slip downwards. On an oceanic mountain, large landslides can sometimes cause the entire top to slide downslope, leading an island to disappear in the process, wholly – as in the case of Mamata and Tolamp – or almost wholly. Examples of the latter, which demonstrate this explanation to be correct, are also known.

In the year 1834, in the waters of what today we know as Papua New Guinea, as part of the crew aboard the decrepit brig *Margaret Oakley* captained by the untrustworthy Benjamin Morrell (see Figure 2.2), Thomas Jefferson Jacobs described 'a conical island, that rises abruptly from the water, and towers more than 300m into the air. It is composed of lava, clay, pumice and rotten stones, sulphur and cinders. Steam issued from its summit and its sides were denuded and washed into deep chasms. I named it Cone Island'.[19] The sketch Jacobs made is shown in Figure 9.3. It is the only image we have of an almost now-vanished island from this time.

Research suggests it was gravity alone, acting on the steep flanks of Cone (later Ritter) Island that caused it to almost disappear.[20] The process had commenced, perhaps

Figure 9.3 *Cone (Ritter) Island in Papua New Guinea, 'a vast columnar portal' drawn and described in 1843 by Thomas Jefferson Jacobs. The sole visible remains of the island today are shown in black.*

even by the time of Jacobs' visit, with a slow slippage of the island's western flank, which culminated in its catastrophic collapse that ripped off the head of the island to expose its neck from the centre of which poured liquid rock. When this came into contact with the cold ocean water there were explosive eruptions. A train of tsunami waves generated by the flank collapse swept over the disappearing island, the only piece of which we can see today, shown in Figure 9.3, being a minute part of that visible in 1843.

These catastrophic events occurred at approximately 6 a.m. on 13 March 1888, something we know because there are abundant eyewitness accounts, fortuitously aided by the fact that New Guinea had just become a German colony and was being systematically inventoried at the time. The collapse powered tsunami waves 20m high that drove within a few minutes across nearby islands like Sakar and Umboi, killing thousands, underscoring the fact that unstable oceanic islands in every part of the world are indeed deserving of our close attention.[21]

Building on this last example, the next chapter considers more broadly in time and space how land collapses, sometimes on a gigantic scale.

CHAPTER TEN

Falling Apart: Collapsing Continents and Islands

M ost of what we today call Norway was smothered by a thick ice sheet during the last great ice age, the crust of the Earth below it depressed by the massive load it was forced to support. Then as the ice started melting about 14,000 years ago, newly relieved of its burden, the underlying crust began to rebound upwards (shown in Figure 8.2B), a process continuing today across most of Scandinavia. The ice sheets started melting along their fringes, exposing land that drew herds of reindeer and elk to feast on the newly grassed hills fringing the pristine pine and spruce forests spreading above the majestic fjords that were forming along Norway's west coast. In the wake of the herds came people, people of the Mesolithic, who hunted these prey animals, but also found unexpected treasures in these new landscapes. Along the Norwegian coast, people discovered 'beach flint', carried northwards within ice floes that later melted, releasing the flint and allowing it to be washed onshore. It is staggering to contemplate that, even though there is no flint in the solid rocks that compose Norway, flints of this kind became the commonest rocks for the manufacture of stone tools here in Mesolithic times, 10,000 years and more ago.

Late autumn, about 8,150 years ago on the west coast of Norway. The people who had spent their summer hunting reindeer and elk in the verdant hills beyond the heads of the fjords had returned to the coast, their usual practice at this time of the year. They were ebullient, their smokehouses filled with river salmon and dried reindeer haunch. They hunkered down in their villages on the lowlands at the heads of the fjords to await the arrival of winter. They carved fishhooks from animal-limb bone and assiduously reconnoitred the locations of nearby colonies of seal and sea-otter that would help sustain them through the dark months ahead. As they had descended the steep cliff paths leading down to the shore, they barely noticed the last of the year's mountain fern moss matting rock faces along the sides of the paths yet, more than 8,000 years later, it is this same unassuming moss that allows us a unique insight into the precise timing of an unprecedented catastrophe – the Störegga Slide.[1]

Störegga means 'great edge' in Old Norse and is a fitting name for a super-sized collapse of the continental shelf of Norway that involved an area of some 95,000km^2 – about the size of the US state of Indiana, a bit larger than Portugal – slipping downslope into deeper water. It left behind a scar so long that the first geologists to map it thought it was simply the wave-trimmed edge of the continental shelf itself. Only later, when giant tongues of debris leading back to the scar were found on the adjacent deep ocean floor did it become obvious that the scar was in fact the headwall of an astonishingly large landslide.

The abrupt change in the form of the ocean floor resulting from the Störegga Slide generated tsunami waves, not readily detectable in deep water but which, as they approached the nearest coasts, gradually became higher and higher. When these waves barrelled along the closest fjords in western Norway and smashed into the settlements clustered around their heads, the largest might have been 12m high. Many people would have died instantly. The survivors, half mad with disbelief, would have struggled to survive the approaching winter because their food reserves would have been decimated, their familiar environments shredded.

After the waves had run ashore, halfway up the back walls of the fjords, they washed back down scraping off the mats of the mountain fern moss that then became incorporated into shell-rich deposits on the fjord floors … where they lie today under 6m of water and mud. Fortuitously preserved both because of the rapidity of their burial and the alkalinity (conducive to chlorophyll preservation) of the sediments that quickly buried them, it has been possible to find pieces of this green moss so well preserved that even the stage of their annual growth can be determined. The fact that the preserved mosses all show they were buried in late autumn tells us that the fjord-head villages were probably full at the time these monster waves hit them.[2]

Investigations have found that there was in fact a number of Störegga slides, one giant one involving the movement of perhaps 3,200km³ of material, followed by a multitude of smaller ones. To understand the hazard posed by such mega collapses along continental margins, it is critical to have an accurate picture of the largest event. And rather than trying to dig through tens of metres of landslide debris 4km below the ocean surface, it proved far easier to work out ages for the associated tsunami deposits and use these to simulate the heights and speeds of the associated waves, from which the size and speed of the Störegga slides can be inferred. The results show that the main slide moved downslope at rates of 25–30m each second with pauses of 15–20 seconds between the movement of each slide block. The water that was rapidly displaced by the biggest slide produced waves that shot out across the ocean in all directions, back across western Norway, reaching the Faeroe Islands, Greenland, and pounding parts of the coast of Scotland and England (Figure 10.1).

Travelling so fast it was like a wall of cement, the largest wave was 25m high when it reached the Shetland Islands,

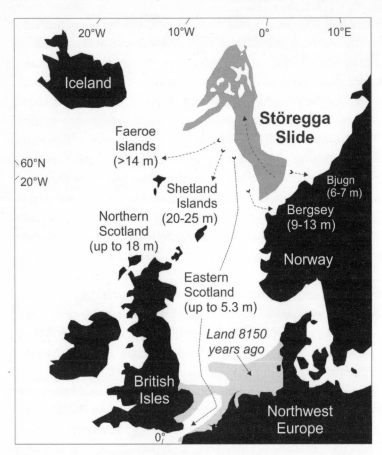

Figure 10.1 *About 8,150 years ago the Störegga Slide occurred off the coast of Norway, creating waves as much as 25m high that smashed into nearby coasts and perhaps severed the land connection between Britain and the rest of Europe.*

as high as 18m when it hit the north coast of Scotland and less than 6m high – still not trivial – as it ran south along its east coast. Evidence has emerged from several of these locations of the memorable impacts the Störegga tsunami had on coastal peoples eight millennia ago.

The evidence from Howick, on the Northumberland coast of north-east England, is mostly negative – a conspicuous gap in the record of the area's human occupation.[3] The gap represents the loss of 3,000 years of the artefact-rich sediments accumulating in a coastal river valley, something explainable by the inrush of the Störegga tsunami that stripped off the upper parts of the valley fill. The tsunami also left behind a tell-tale signature – a layer of large-sized (coarse) sand that could only have been lain down in a single rapid event. While the tsunami must have had a huge impact on the people of the Howick area, the precise nature of this is uncertain. More complete examples of the human impacts of the Störegga tsunami are known from Scotland, where stone tools from coastal settlements of hunter-gatherers dating from this time are scattered across the landscape. Probably the settlements and their occupants too.

One of the most intriguing suggestions concerning the effects of the Störegga tsunami is that which envisages it transforming Britain into an island, maybe literally overnight.[4] During the last ice age, the British Isles were joined to Europe in several places, but then as sea level rose after the end of the ice age these land bridges became progressively narrower and were finally severed. The last to be cut was that connecting East Anglia to the north-west Europe coastal plain (shown in Figure 10.1), an event that occurred around the same time as the Störegga tsunami when the ocean surface here was some 14m lower than it is today. The tsunami waves swept over this land bridge, perhaps cutting it temporarily but, more importantly, signalling its vulnerability to its inhabitants and foreshadowing its imminent submergence by rising sea level.

The Störegga Slide was caused when the thick pile of sediments laid across the continental shelf in this part of north-west Europe suddenly became unstable and collapsed. This is thought to have happened for one of two reasons and the precise time when the slide occurred provides important clues.

The presence of the remains of green moss, rapidly killed, in the Störegga tsunami deposits have allowed the age of the Slide to be fixed to a time around 8,150 years ago, which just happens to be 'the chilliest part of the greatest cold snap of the last 10,000 years', the so-called 8,200-year cold event.[5] This event is believed to have been initiated when in eastern Canada about 8,500 years ago a large ice-dammed lake, the same implicated in the earlier interruption to the Neolithic Revolution (discussed in Chapter 6), suddenly burst through its ice bonds and poured into Hudson Bay and thence the North Atlantic Ocean.[6] By abruptly introducing a mass of freshwater, the thermohaline circulation in the North Atlantic — which keeps its high-latitude shores warmer than they otherwise would be — shut down, plunging north-west Europe into around 100 years of bitterly cold conditions. The coincidence between the Störegga Slide and the 8,200-year cold event suggests their causes are connected, but it is not clear how.

So, what did actually trigger the Störegga Slide? One explanation involves the huge quantities of terrestrial debris — soil and fragments of rock — washed off the land surface of northern Europe by melting ice as the world warmed in the aftermath of the ice age. This debris accumulated in offshore areas like the Norwegian continental shelf where the Störegga Slide originated. Normally, when sediment accumulates on the ocean floor in such locations, the water within the sediment — actually filling the pores between the individual particles — is progressively squeezed out as more and more sediment is slowly piled on top. But it is possible that ice was melting and sediment was accumulating so rapidly at this time on the Norwegian continental shelf that there was insufficient time for all the water to be completely squeezed out of the pores. As a result, the sediment pile became buoyant — it began floating on layers of saturated sediment within it — which eventually led it to collapse.

A second possible explanation involves gas hydrates, which accumulate in layers within thick piles of sediment such as those that blanket continental shelves. This is not too difficult to understand. The upper parts of the oceans are full of microscopic carbon-based organisms (plankton) whose remains, when they die, sink to the ocean floor. Although they may be buried there for millions of years, eventually bacteria get to work on them and release methane gas. Pressure forces the methane upwards from places where it is comparatively warm (beneath several hundred metres of ocean-floor sediment) to places nearer the ocean floor where it is much colder. So cold is it here that ocean water will sometimes freeze and form ice cages around handfuls of methane gas, producing what are known as gas hydrates, which exist in layers within sediment piles such as those off the Norwegian shelf.

Now sometimes these gas hydrates – ice-encaged lumps of methane – break up. As a result, the gas escapes and the sediment body of which they were once an integral part abruptly becomes unstable and may collapse. This scenario may be what initiated the Störegga Slide. Perhaps the sea level rose and increased the pressure (the weight of overlying water) on the gas hydrate layers causing them to break up.

During the time that modern humans have strode the Earth, almost 200,000 years, mega collapses of the continental shelf like Störegga have been rare. But the Earth is much older, and the geological record shows us clearly that an event the size of the Störegga Slide is far from without precedent.

River deltas are not to everyone's taste. Vast, featureless, flat, damp areas often covered with dense impenetrable forest alive with hematophagic insects. Yet the one common factor that draws humans to deltas in every part of the world is the soil. Having been washed off land surfaces upstream, it is exceptionally fertile and of course plentifully watered. Crops that grow in delta areas are often the envy of farmers elsewhere.

But apart from the insects and the periodic floods (which introduce the topsoil), there are less common hazards that affect delta life. One is that deltas are liable to sink because the sediments from which they are built become compacted – the great weight of the largest deltas may even change the shape of the underlying Earth's crust. But a delta can also sink when its sediment pile suddenly becomes unstable and slumps. We saw this in Chapter 3 where sudden slumping at the front of the Nile Delta is implicated in the disappearance of the cities of Herakleion and Eastern Canopus. The Amazon, another uncommonly long river, does not appear to have a delta at its mouth, at least not one that emerges above the ocean surface.

The earliest written account we have of the Amazon River dates from around the end of January in the year 1500 when Vicente Añes Pinçon 'entered a tract of ocean where the water was fresh at a distance of thirty leagues [145km] from the coast'.[7] Realising this must be a river mouth – the sailors could not see to confirm this from so far away – Pinçon sailed west and eventually entered a river channel he named Rio Santa Maria de la Mar Dulce, later rechristened the Amazon.[8]

Where a delta is absent from the mouth of a sizeable river, we might wonder what then happens to all the sediment it transports to the ocean – and in the case of the Amazon there is quite a lot of it. Measurements at Óbidos (Brazil) in the 1960s and 1970s found that the Amazon carried between 800–900 million tonnes of sediment to the sea every year. Since then this figure has risen, largely as a result of accelerating deforestation in the Amazonia lowlands, to between 1–1.2 billion tonnes. That's more than 2,000 tonnes of South America being dumped in the South Atlantic Ocean every minute!

The absence of an emergent delta in such a situation can be explained by the absence of any suitable foundations on which river sediments might accumulate, something that clearly applies to the mouth of the Amazon River, seaward of which the ocean floor drops steeply. But the absence of a delta can also be explained by how fast sediment is carried away from the river mouth. Pinçon had an alarming experience of this after his ships anchored in the lowermost Amazon River more than 500 years ago.

> As the ships lay at anchor, a fearful noise was made by the strong impact of the fresh water against the water of the sea coming to meet it and the ships were raised four fathoms [7.3m] and from this proceeded great danger.[9]

Pinçon experienced what the people of the Amazon call *pororoca*, a tidal bore, in which on the rising tide – literally with a roar – seawater builds into a wave up to 4m high that rushes upriver at speeds as fast as 25km per hour. The existence of the *pororoca* underscores the power of the ocean at the mouth of the Amazon and explains its efficiency at removing the huge quantities of sediment that come out of it.

Drilling into the deep-ocean floor off the mouth of the Amazon Delta has found much of the sediment to have been washed off this part of South America over the past few million years. And while it is unlikely there was ever a large emergent delta at the mouth of the Amazon, it appears that shallow-water deltas developed here on numerous occasions in the past … then collapsed.

Large-scale collapses of deltas do not occur very often, thankfully. In fact, it is fair to say that rather than eyewitness accounts, the only reason we know they occur at all is because of the presence of giant ocean-floor sediment fans. One of the largest such fans ever mapped lies at the mouth of the Amazon River and research shows it formed from not just one catastrophic collapse but several. If this sounds vague, well it is. But it is also an acceptable conclusion when you are dealing with a feature beneath almost 2km of ocean water comprising a number of individual slide features each the size of Jamaica.[10]

Analyses of the sediments in the deep-water Amazon Fan show they travelled from the collapsing delta more than 200km out to sea. The organisation of the sediments in the fan – what is termed its architecture – shows they were deposited rapidly, during one or more catastrophic failures of the delta front.

So, what might have caused these failures? There are two possibilities. Investigations show that two of the fan's constituent slides occurred when sea level was falling rapidly 35,000–45,000 years ago and gas hydrate break-up appears the likeliest explanation. Yet two other slides happened more recently, around 13,000 years ago, when sea level was rising after the last ice age ended. Gas hydrate break-up may again seem the obvious culprit,[11] but the scientists who studied these younger collapses came to favour another. This involves the effects of post-glacial warming on the extensive ice masses that covered the highest parts of the Andes Mountains, where the Amazon River rises, during this ice age. Rising temperatures melted much of this ice, the meltwater produced pouring into the Amazon headwaters, racing downstream and out across the delta, a wobbly pile of waterlogged sediment, subjecting it to uncommon mechanical stress and leading it to collapse.

People may not have been present in South America 35,000 years ago (although there is evidence to the contrary[12]) when the first-known mega collapses of the

Amazon Delta took place, but they were certainly there 13,000 years ago when the more recent ones occurred. So, imagine the effects of Andean meltwater flooding on the people in the Amazon lowlands, even those living on the now-drowned Delta. We know nothing of them or the places they called home simply because all traces have been washed out to sea. This illustrates what science has slowly grown to realise, particularly over the past few decades, about how much of our history is no longer visible to us, like the missing three millennia at Howick mentioned earlier, like the collapsed deltas of the Amazon.

Just because much of human history is invisible does not mean it was never there or that its existence was unimportant. In order to properly understand ourselves and our journey as a species, our challenge is to acknowledge the existence of this hidden history and try, iteratively and painstakingly, to piece it together from the fragments we can see. Although he was thinking about geological rather than human history, Charles Darwin captured the point well in *The Origin of Species*.

> I look at the natural geological record, as a history of the world imperfectly kept, and written in a changing dialect; of this history we possess the last volume alone, relating only to two or three countries. Of this volume, only here and there a short chapter has been preserved; and of each page, only here and there a few lines.

Other people, particularly those knowledgeable about non-written histories passed down from their ancestors, have known of this for a long time. The point was evocatively made by T. S. Eliot in his 1942 poem *Little Gidding*.

> Ash on an old man's sleeve
> Is all the ash the burnt roses leave.

Dust in the air suspended
Marks the place where a story ended.

To my mind, some of the scenically most stunning areas of the western United States lie along the unspoilt windswept coasts of Oregon and Washington states. Offshore lies an ocean trench, marking the Cascadia subduction zone, where the Juan de Fuca crustal plate is being thrust eastwards and down below its North American counterpart. The Cascade Mountains are a chain of mostly active volcanoes that are an on-land expression of this subterranean thrusting, being pushed upwards as the Juan de Fuca Plate pushes ever further down beneath them.

Being wholly undersea, the surface of the Juan de Fuca Plate is draped with great thicknesses of ocean-floor sediment, largely the products of millions of years of erosion of adjacent North America. As the Juan de Fuca Plate pushes downwards, so this sediment is scraped off its surface to form a wedge-shaped pile of intensely folded sediments. This sediment wedge is inherently unstable, sliced every which way by faults, ceaselessly compressed and periodically rattled by the earthquakes that characterise such dynamic settings.

Small wonder then that parts of this wedge occasionally collapse, although the evidence is not always immediately apparent, given the great water depths in which it lies. The many areas of hummocky terrain on the sea floor here first alerted researchers to the possibility that ancient landslides lay buried beneath younger sediments. The hunch was confirmed through use of sub-surface imaging techniques like side-scan sonar, which allowed buried landslide lobes to be identified. What researchers were not prepared for was the size of these landslides, in recognition of which they were dubbed super-scale failures.[13]

The submerged city of Ys may lie somewhere within the Baie de Douarnenez or the Baie des Trépassés. It is told that Dahut, Princess of Ys, opened the floodgates one day letting water flood the city, forcing its abandonment. The painting below shows Dahut fleeing with her father, King Gradlon, but being pulled from his horse by St Guénolé, who knew what she had done.

Left: The submerged city of Pavlopetri existed 5,000 years ago and involved streets and double-storey buildings with courtyards.

Left: Image showing cart tracks that may date from about 2,000 years ago when the isthmus between Pavlopetri and Elaphonisos Island (in the distance) was only slightly submerged.

Below: The reconstructed head of Mana (left), a woman who lived almost 3,000 years ago in Fiji – one of these islands' earliest settlers. Her stained teeth in the skull (right) suggest she chewed kava, a stimulant that helped keep people alert on the lengthy ocean voyages that marked the initial settlement of western Pacific Island groups. Such a process may have been driven by the drowning of densely-populated coastal lands along the fringes of larger landmasses to the west.

A

VOYAGE
TO
St. KILDA.

The remoteſt of all the *Hebrides*, or Weſtern Iſles of *Scotland*:

GIVING

An ACCOUNT of the very remarkable Inhabitants of that Place, their Beauty and fingular Chaſtity (Fornication and Adultery being unknown among them); their Genius for Poetry, Muſic, Dancing; their furpriſing Dexterity in climbing the Rocks, and Walls of Houſes; Diverfions, Habit, Food, Language, Diſeaſes and Methods of Cure; their extenſive Charity; their Contempt of Gold and Silver, as below the Dignity of Human Nature; their Religious Ceremonies, Notion of Spirits and Viſions, &c. &c.

To which is added,

An ACCOUNT of *Roderick*, the late Impoſtor there, pretending to be ſent by St. *John Baptiſt* with new Revelations and Diſcoveries; his Diabolical Inventions, Attempts upon the Women, &c.

BY M. MARTIN, GENT.
The FOURTH EDITION, correćted.

The Inhabitants of St. Kilda are almoſt the only People in the World who feel the Sweetneſs of true Liberty; what the Condition of the People in the Golden Age is feigned to be, that theirs really is. P. 67.

LONDON:
Printed for DAN. BROWNE, without *Temple-Bar*; and LOCKYER DAVIS, in *Fleet-Street*.

MDCCLIII.

---

[15]

above four Inches long, and about two in Diameter, each Piece Sexangular.

Upon the Weſt fide of this Iſle lies a Valley with a Declination towards the Sea, with a Rivulet running through the middle of it, on each fide of which is an Afcent of half a Mile; all which Piece of Ground is called by the Inhabitants, The Female Warrior's *Glen:* This *Amazon* is famous in their Traditions: Her Houſe or Dairy of Stone is yet extant; fome of the Inhabitants dwell in it all Summer, though it be fome Hundred Years old; the whole is built of Stone, without any Wood, Lime, Earth, or Mortar to cement it, and is in form of a Circle Pyramid-wiſe towards the Top with a Vent in it, the Fire being always in the Centre of the Floor; the Stones are long and thin, which fupplies the Defećt of Wood: The Body of this Houſe contains not above Nine Perfons fitting; there are three Beds or low Vaults at the fide of the Wall, which contains five Men each, and are feparated by a Pillar; at the Entry to one of thefe low Vaults is a Stone ſtanding upon one end; upon this ſhe is reported ordinarily to have laid her Helmet; there are two Stones on the other fide, upon which ſhe is faid to have laid her Sword: they tell you ſhe was much addićted to Hunting, and that in her Days all the Space betwixt this Iſle and that of *Harries,* was one continued Tract of Dry Land. Some years ago a Pair of large Deers-horns were found in the Top of *Oterveaul* Hill, almoſt a Foot under Ground, and a Wooden Diſh full of Deer's Greafe. 'Tis faid of this Warrior, that ſhe let loofe her Grey-hounds after the Deer in *St. Kilda,* making their Courfe towards the oppofite Iſles. There

C 2   are

Above: The Scottish islands of Harris and St Kilda may once have been one. Excerpts from this 1753 edition of Martin's *A Voyage to St Kilda* include an account of the 'Amazon' said to be addicted to hunting at a time when the ocean floor between St Kilda and Harris islands was dry land.

Right: A 1590 depiction of a Pictish warrior woman by Theodor de Bry.

Right: Claims that human-made structures, even a submerged 'city', lie here off the coast of Yonaguni Island (Japan) ignore onland geology which shows that nature creates such structures.

Below: The natural rock archway named *Li'anga Huo A Maui* on 'Eua Island in the islands of Tonga is said to have formed when the demigod Maui pulled his hoe out of the rock into which his mother had thrown it. Along the shoreline here is an ancient volcanic basement tens of millions of years older than most Pacific Islands, evidence that 'Eua is a fragment of continental crust that became stranded here long before other islands in the region formed. Luo, Zofia Kielan-Jaworowska and Richard Cifelli at Kielan-Jaworowska's home.

Right: The memorable 1883 eruption of Krakatau (Krakatoa). A newspaper at the time showed a sketch of the island that was 'said to have disappeared'.

Right: The island of Kehpara, Pohnpei, Federated States of Micronesia, the inhabitants of which have stories about neighbouring islands which disappeared.

Right: View along the Palouse Canyon in the Scablands of Washington State, cut during a series of mega-floods that followed the last ice age.

Right: In 1811, Canadian cartographer David Thompson recognised that the Wallula Gap in Washington had a catastrophic origin. He considered that 'the finger of the Deity has opened by immediate operation the passage of this river through … solid materials'.

Above: The El Golfo landslide scar on Hierro (Canary Islands) formed after a massive island flank collapse 15,000 years ago.

Below: Taioha'e Bay, Nuku Hiva Island (Marquesas) formed when one of this island's sides collapsed.

Below: The *pali* (landslide scar, meaning 'cliff' in Hawaiian) from Nu'uanu overlook on the Hawaiian island of Oahu.

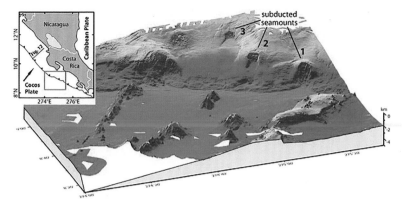

Above: Image of the ocean floor off the coast of Costa Rica (Central America) showing seamounts burrowing into the continental crust; several more seamounts are heading in the same direction.

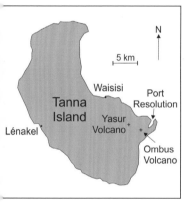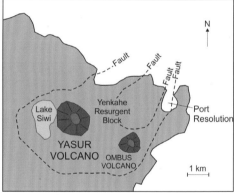

Above and below: The island of Tanna, Vanuatu, Southwest Pacific, is home to Yasur – one of the most active volcanoes in the Pacific. The easternmost part of Tanna, which houses Yasur and its younger sibling Ombus, forms a separate structural block that shifts in response to movements of magma within the underlying crust. The western shore of Port Resolution harbour, below, was raised ten metres in 1878.

Left: Around 1,500 years ago, St Brandan sailed from Ireland into the unknown Atlantic Ocean, reportedly discovering many islands including one that turned out to be a giant fish.

Lines of standing stones (menhirs) around Carnac, France.
Above: The Ménec alignments can be seen. **Inset:** A modestly sized menhir.
Below: The partly-submerged line of standing stones on the coast of the island of Er Lannic, in the Gulf of Morbihan, Brittany.

Three such slides were eventually mapped here – Heceta, Coos Basin and Blanco. Collectively they involved the downslope movement of almost 16,000km^3 of the sediment wedge. The youngest (Heceta) occurred about 110,000 years ago, the oldest (Blanco) around 1.2 million years ago. All occurred in deep water, well below the level at which gas hydrates are liable to break up, so this cannot be the reason for their collapse. Rather it seems each failure was triggered by a large-magnitude earthquake.

In contrast to the western expanses of the adjoining Sahara Desert in tropical Africa, the coastal fringe of Mauritania is verdant and well-watered, and where most of its people live today – but this was not always the case. This great desert was not always as extensive or continent-spanning as it is now. In times past, inland cities sustained by the productivity of surrounding lands and their importance to Berber trade networks were dotted across the country. They include Chinguetti, Ouadane and Tichitt, each inscribed as World Heritage Sites in 1996. Ancient rock paintings at Agrour Amogjar near Chinguetti depict giraffes and lions, even cattle, occupying its once fertile well-watered landscape. Driven by increasing dryness and vegetation loss, the Sahara has expanded forcing most of its former residents towards the ocean.[14]

Only in the past 40 years has the ocean floor off most of the coast of Africa been systematically surveyed and there are many parts that remain comparatively poorly known, but where interest – driven by the possibility of discovering economic reserves of hydrocarbons – is burgeoning. One such area is off the coast of Mauritania. Here, under some 100,000km^2 of the seabed, economically viable deposits of hydrocarbons have been discovered, clustered on the continental shelf several tens of kilometres west of the capital Nouakchott. Mapping of the continental shelf here has also revealed many ocean-floor landslides, collectively named the Mauritania Slide Complex. It is obviously worth knowing something about the history of this slide complex, not just because landsliding might interfere with ocean-floor drilling activities, but also because any future such landslides could also generate large waves that could wash back over the low-lying coastline to the east.

The Mauritania Slide Complex covers about 34,000km^2. What is probably the youngest slide occurred a mere 10,000 years ago with several older slides dating to as much as half a million years ago. With no evidence for gas hydrate accumulation in this area, offshore landsliding here is thought to be linked to local oceanographic processes. It is an intriguing story. Coastal winds in this part of the east Atlantic Ocean (that is, off Mauritania as well as Portugal further north) blow southwards parallel to the coast. The effect of the Earth's rotation on this is to divert those winds slightly westward, the upshot of which is that surface ocean waters off these coasts are moved away from them. This in turn allows deep-ocean water to rise to the surface, a phenomenon known as upwelling.

Upwelling is good news for coastal people because it brings nutrients from deep water up to the surface. And surface concentrations of nutrients spawn a richness of ocean-surface creatures (fish are the ones of most interest to people) and those that feed on them (such as seabirds). Less visible yet abundant in areas of upwelling are the concentrations of microscopic organisms that multiply here for the same reasons as larger life forms. The death of these myriad microscopic organisms in upwelling areas, such as the continental shelf off Mauritania, leads to the accumulation of their remains as organic sediment on the ocean floor. And the unusually high accumulation rate of organic sediments here may be the main reason for the landsliding. For as we saw in reference to the post-glacial Amazon Fans, uncommonly high rates of offshore sediment accumulation may not leave enough time for pore water to be squeezed out of a sediment pile, rendering it unstable and liable to collapse. Like a soggy layer cake, made too high.

Dust has been a feature of daily life in coastal Mauritania forever it seems. Today it is ubiquitous, all-intrusive, a constant reminder of the proximity of the world's greatest sand desert and its expansionist tendencies. Great dust storms – *haboob* – are periodically whipped up by strong winds blowing across the western Sahara, forcing the inhabitants of countries like Mauritania to shelter until the storms pass. There are long-term cycles of dust accumulation here consisting of alternations between normal conditions, as appear to exist at present, and shorter periods, lasting decades, when higher-than-average amounts of windblown sand have been dumped on the Mauritanian coast and offshore. The existence of these shorter periods of more-intense dust deposition has been discovered only recently from drilling into sediments on the Mauritanian continental shelf. What makes these periods so interesting is that each of them – and at least five occurred within the past 5,000 years – seems to have preceded an episode of ocean-floor landsliding.[15]

The occasional strengthening of the trade winds has been mooted as the most likely suspect for these periods. At such times, these winds, stronger than usual, pick up Saharan dust, carry it west and drop it off the West African coast, further west than normal, where ocean currents drive it up the underwater gullies cut into the continental slope. The rapid build-up of sediment at the heads of these gullies creates enough pressure to cause their entire sedimentary fill to mobilise and slide into deeper water.

If this is correct, then it is an intriguing example of a connection between climate variations and sea-floor landsliding.

Saharan dust is not confined to North Africa. Sometimes it finds its way across the entire Atlantic Ocean to the islands of the Caribbean, occasionally even Florida. While implicated in respiratory problems for people here, uncommonly dusty years also coincide with years when fewer tropical cyclones – hurricanes as they are called in the Caribbean – develop in this region. And on the other side of the United States, research shows that winter snowfall increases in the Sierra Nevada mountains of California whenever there is an abundance of Saharan dust in the atmosphere.[16]

To the casual observer it may appear that the flanks of the continents where most of us live are innately and frequently unstable, but that is not the case. Like many uncommon and extreme events, we sometimes allow our insecurities, even our fatalism, to exaggerate the threats these pose. But there are places where these are greater than average. Places like the steep-sided islands in the middle of our oceans, collapses of the flanks of which may sometimes occur.

Many years ago, it is said, the god named Uoke roamed from island to island in the central Pacific carrying a giant crowbar that he used to split entire islands apart, breaking them into pieces that he tossed into the ocean where they disappeared. By all accounts, Uoke was a grim individual, quite a contrast to many other gods in the Polynesian pantheon like Tangaroa and Hina, the sun and the moon, and Maui the mischief-maker who, as noted in Chapter 2, fished up islands for people to live on.

Arriving in 1935 on remote Easter Island (Rapa Nui) in the south-east Pacific Ocean, a Capuchin friar, Father Sebastian Englert, spent 35 years there as parish priest and collected many stories – including several about Uoke – from the islanders. They explained that their ancestors had come from a place called Hiva, perhaps an island (like Hiva Oa or Fatu Hiva) in the Marquesas Islands of French Polynesia, where Uoke had done great damage. The Easter Islanders told Father Sebastian that Uoke eventually caused such a *cataclismo* there that their ancestors had to abandon Hiva and sail to uninhabited Easter Island 3,700km away. But Uoke followed them! And in typical fashion, after he arrived, he went around systematically prising bits off the island's cliffs – you can see the scalloped coast made by Uoke's crowbar when coming into land at Mataveri Airport – until on the unusually hard rocks at a place called Puko Puhipuhi his crowbar broke and Uoke passed into oblivion.[17]

A great story you might think, but it is more than that. Like many similar stories, it echoes people's attempts to rationalise memorable observations, in this case the occasional cataclysmic collapse of the flanks of steep-sided islands.[18] And both Easter Island and most of those in the Marquesas show signs of many such collapses.

Located well within warm tropical ocean waters, the islands of the Marquesas are singular because there are no coral reefs fringing their shores, as there are in every other tropical Pacific Island group. The likeliest explanation for this is the islands' flanks are so steep that there is simply no foothold on which reefs could develop. Add to this the fact that these flanks are so inherently unstable that, even should a reef start to grow, the chances are it would soon find itself in deep water once more.

Fatu Huku is an island in the Marquesas, home today only to vast numbers of nesting seabirds – boobies, frigate birds and terns. In terms of size, Fatu Huku is today a fairly insignificant member of the group, having an area of just some 1km² despite rising almost 400m above sea level. The mysteries about Fatu Huku are several. For while no islands in the Marquesas have any coral reefs developed off their coasts today, Fatu Huku has fossilised reef on its summit plateau. A puzzle indeed, and one acknowledged in Marquesan legends that refer to Fatu Huku as 'an island turned upside down', an allusion to the reef being on the mountain top rather than off the coast as is normal. And then there are the ancient shrines and altars (*marae*) on Fatu Huku, similar to many found on larger islands in the Marquesas yet oddly numerous on this small lump of rock. Maybe Fatu Huku had special significance in the religious beliefs of ancient Marquesans, but that suggestion rings a bit hollow when you consider this island is neither nowhere near as high as the larger islands nor is it located centrally within this island group. An improbable candidate for a site of high spiritual significance.

So, maybe with Fatu Huku we are dealing with an island that was actually once much larger than it is today. An incredible suggestion? Elsewhere perhaps, but not in the Marquesas where there is a long history of large island-flank collapses. And not in the case of Fatu Huku, for which there is actually a paper trail suggesting that a large part of the island really did collapse and disappear about 200 years ago.[19]

No slouch when it came to mapmaking, Captain James Cook visited the Marquesas in 1774 and made a chart of the area showing a sizeable island named Hood's Island exactly where the tiny island of Fatu Huku exists today. The same is true of Lieutenant Richard Hergest, en route from the Falkland Islands to Hawaii in March 1792.[20] It seems clear that Fatu Huku was a larger island in the late eighteenth century.

The next map made of this part of the Marquesas shows otherwise. A map by Captain David Porter in 1820 – and all subsequent maps – shows Fatu Huku as it is today: comparatively small, unusually high. The inescapable conclusion is that between 1792 and 1820 something happened to Fatu Huku that reduced it in size by perhaps 90 per cent. A catastrophic flank collapse, comparable to that shown in Figure 9.3, fits the bill.

If all that was not enough, well, there is an oral tradition that may recall the actual event. In *Te Fenua 'Enata* (the Land of Men), as the Marquesas was once called in Polynesia, there is a tradition stating that the former island of Fatu Huku rested on an underwater pillar of rock, protected from dislodgement by a guardian shark.[21] But one day the shark, angered, started thrashing in the water, repeatedly hitting the rock pillar with its caudal fin until the island above toppled over. This story explains this is why coral is now found on top of the present island, itself a fragment of the large island that once existed here.

Since collapses like that at Fatu Huku involve the loss of material, it is difficult to calculate how often they occur. So, with the Marquesas, just as for the Störegga Slide discussed earlier, one approach has been to examine the deposits left behind when giant waves from these collapses washed back across the islands. A clever thought for sure: date the deposits and work out how often big waves from flank collapses strike. Unfortunately, this bright idea has not really succeeded in the Marquesas for most of the giant waves that strike these islands originate much further away, places like Alaska and the Cascadia margin of the western United States where tsunamigenic earthquakes periodically occur.[22]

Yet local island-flank collapses remain a threat in the Marquesas. The people of the coastal town of Omoa on the island of Fatu Hiva will not quickly forget the events of 13

September 1999. It was after lunch, classes had just resumed at the small school when a 7m-high wave tore onshore, demolishing a reinforced concrete wall and pouring through the classroom windows. The headteacher had seen this monster approaching and the children were already climbing out of the windows on the opposite side of the classroom when it struck – no one was hurt.

This tsunami was traced to a cliff collapse 3km away and is estimated to have involved movement of some 5 million m³ of rock. What is especially interesting about this landslide is that it was not triggered by an earthquake – earthquakes are comparatively rare in the Marquesas. Nor, like the similar well-studied Hilina Slump in Hawai'i, did it follow days of heavy rain.[23] The scientists who investigated concluded that it was actually a result of the months of drought preceding it. Layers of clay within the rocks forming the cliff had shrunk as a result of dewatering, causing the entire structure to become unstable and abruptly collapse.[24]

Another group of high volcanic islands with a history of collapse and, a little concerningly, some expectation of future collapse is the Canary Islands, part of Spain, in the eastern Atlantic Ocean. Research into flank collapses of the Canary Islands volcanoes has taught us plenty about the long-term evolution of oceanic islands as well as the threats their occasional collapses pose to coastlines within the same ocean basin. In the case of the Canaries, island collapse is linked to the growth of stratovolcanoes, naturally steep-sided edifices. An eruption here, the swelling of an underground magma chamber there – such factors can lead a stratovolcano to become unstable.

Many examples of rapid island-flank collapse in the Canaries have been described (retrospectively). One of the earliest occurred some 15,000 years ago on the flanks of Hierro Island and comprised three separate failures. First, the north side of Hierro was affected by the El Golfo debris avalanche, which incorporated blocks of volcanic rock as much as 1.2km in diameter. Second, below the debris avalanche in water about 4km deep, a debris flow affected the north side of the island edifice, carrying some 250–350km^3 of material 600km across the deep-ocean floor. Thirdly, there is evidence that a turbidity current developed at the same time and ran downslope way past the end of the debris flow and out across the Madeira Abyssal Plain. The Hierro study excited marine geologists because it showed that different forms of mass wasting (debris avalanches, debris flows, turbidity currents) could all be generated by the same catastrophic event.[25]

The most talked-about flank failure in the Canary Islands has not yet happened (see Chapter 13). During its 24 June 1949 eruption, an estimated 200km^3 of the Cumbre Vieja volcano on the southern flank of La Palma Island dropped vertically 4m – and then stopped. It is likely this collapse will restart during some future eruption of Cumbre Vieja, currently the most active volcano in the Canary Islands group.

Studies of the older volcanic islands in the Canary group show they all passed through a shield stage (when the form of the island resembles an upturned shield), during which collapses are most common, before stabilising in what is called their post-erosional phase. The best-studied example is of Fuerteventura where the ancient submarine volcano (around which the island grew subsequently) has been disembowelled, its innards exposed largely through mega-landsliding. The internal plumbing seems to have been responsible for instigating this. Bodies of liquid rock (magma) pushed upwards inside the volcano break up as they approach its summit, pushing out sideways along lines of pre-existing weakness between older rock layers, forcing these apart and often causing collapse. This in turn can create lines of weakness that did not exist before and these may be targeted by rising magma leading to further collapse in the same place.[26] Insights into these kinds of processes and their obvious practical implications have stemmed from studies of another group of high volcanic islands – Hawaii.

Probably no group of islands has received more attention from geologists than Hawaii. All Hawaiian islands rise from the underwater Hawaiian Ridge, one of the steepest-sided structures on our planet, and their coasts are marked by giant scallops – named *pali* by the first Hawaiians – that mark places where huge chunks of the land once slipped away. Or where Uoke once prised them loose. When these *pali* were first mapped, no scientist could readily credit the extraordinary size of the collapses that created them. Take the Nu'uanu Slide that occurred around 2 million years ago and was responsible for removing the northeastern fifth of O'ahu Island, some 5,000km^3 of volcanic rock. Today the Nu'uanu Pali towers over the town of Kane'ohe and its wondrous coral-filled bay that nestles within the bottom of the half-broken cup left behind by this huge slide. The other side of O'ahu is also marked by a massive flank collapse, the Wai'anae Slide, far older than the Nu'uanu Slide.

Studies of the Hawaiian Islands have focused not only on the on-land evidence for flank collapse, but also on the ocean-floor deposits produced by particular collapses. As you can see in Figure 10.2 (which cannot even show all the sea floor landslides), there has been a prodigious number of mega-landslides affecting the Hawaiian Islands, ever since they first poked their heads above the surface of the Pacific

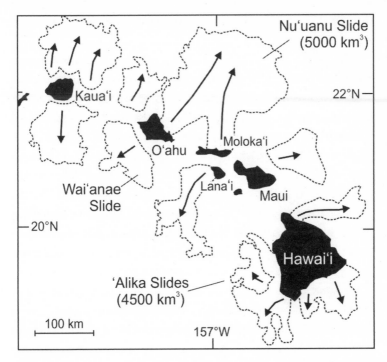

Figure 10.2 *On a geological timescale, the islands of Hawaii are falling to pieces, as shown by the numbers of undersea landslides resulting from island-flank collapses.*

82 million years or so ago. Fortunately these landslides do not occur so frequently that humans, who may have lived in Hawaii for 1,300 years, have any memory of these – although it is comforting nonetheless to know that the islands' stability is closely monitored.

Each of these mega-landslides abruptly displaced a huge volume of ocean water, generating giant waves that dispersed across the Pacific but also washed back across the Hawaiian Islands. On islands like Lana'i and Molaka'i you can see wave-deposited corals high above the present coast, from which it has been inferred that some of these waves reached 200m in height. So incredible was this inference for geologists to process, you will not be surprised to hear that there was considerable resistance to it at first – as there still is.

Harold Stearns was a tall, thin, no-nonsense Scotsman who first studied the geology of Lana'i island in the 1930s where he saw what he interpreted to be an ancient wave-cut shoreline more than 200m above the present one.[27] He thought the ocean surface here must once have been this high, for how else could a shoreline have formed there? The discovery in the 1980s of numerous submarine landslides on the underwater flanks of the Hawaiian Islands provided the impetus for supposing that it was in fact very large waves that had carved this and other high-level shorelines on the islands themselves – the so-called Giant Wave Hypothesis.

At first glance, the connection between giant waves and high-level shorelines seems uncontroversial. Here is a probable scenario. One day, loosened by an earthquake or during prolonged heavy rain, a massive section of one of the Hawaiian volcanoes is detached and slips down the island's flanks, displacing a huge mass of ocean water, some of which then builds up to a great height and rushes outwards from the slip site. The giant wave – a mega-tsunami – crashes into the side of a nearby island, ripping up near-shore coral reefs, rushing inland, driving upslope and carving notches in valley headwalls, eventually dropping the material it is carrying, including coral fragments, high on the island's flank as it loses power. The water recedes after a short time, but the gravels and the landforms remain – an eternal reminder of this extraordinary event.

Later research by sedimentologists like Anne Felton found that most of these high-level gravels, actually no higher than 190m above sea level, show no clear signs of having been deposited in a high-energy situation, as would have been the case had they been laid down by a giant wave. Rather, they appear typical of gravels found along modern coastlines in

Hawaii.[28] This interpretation invalidated the Giant Wave Hypothesis. So how *could* a shoreline have formed 190m above sea level on such an island?

The only answer is that the level of the island itself changed. In the case of Lana'i Island by almost 200m in a million years. That is a big ask, especially in the ocean basins. Usual rates of island uplift are far less, but tectonic geologists like Barbara Keating have made a special plea for Lana'i because it seems to have risen up the side of a bulge in the ocean floor (a crustal flexure) within this time period.[29] So is that the end of the story? Lana'i is one island that has shot upwards. Right?

Perhaps not. For like an attentive parent who ensures their child completes a task fully, so science continues to provide a forum for scientists to doubt until all available evidence is in and a final judgement can be made. So, just when it seemed that the unanimous verdict on the high-level shorelines on Lana'i had been delivered, new research raised major doubts about it. This research shows quite compellingly that Lana'i could not have risen anywhere close to the amount needed to for these high-level gravels to have formed along a shoreline 190m above present sea level. In fact, it is just as likely that Lana'i has been sinking, at least for the past 30,000 years or so. The researchers conclude that the Giant Wave Hypothesis for the high-level gravels on Lana'i is the correct one after all.[30]

Such a conclusion may appear irksome for those whose careful studies led them to an opposing view, but it does now seem likeliest (in my impartial view) that mega-tsunamis have indeed cut shorelines and dumped reefal materials high in the Hawaiian islands from time to time. Corroborative evidence comes from the eponymous island of Hawai'i, also known as the Big Island, the youngest in the group and one that has certainly never moved upwards. A fossil-rich marine conglomerate reaching 61m above sea level is found on the Kohala coast of Hawai'i Island. This conglomerate is believed to have been deposited by a wave an incredible 400m high washing over this coast and reaching 6km inland of it. The conglomerate dates to 110,000 years ago, making it likely that the wave responsible was one produced by the giant 'Alika 2 landslide from nearby Mauna Loa volcano, shown in Figure 10.2.

Even such massive waves did not remove any of the islands they encountered although, as we shall see in the next chapter, extreme waves can be held responsible for the disappearance of some coastal lands and even – in special circumstances – entire islands.

'Huge and Mighty Hilles of Water': Monstrous Waves

It was a fine clear morning in western England on 30 January 1607. The inhabitants of the coastal lowlands fringing the Bristol Channel had no reason to suppose disaster was imminent. But then,

> ... about nine of the morning, the same being most fayrely and brightly spred, many of the inhabitants of these countreys prepared themselves to their affayres then they might see and perceive afar off as it were in the element huge and mighty hilles of water tombling over one another in such sort as if the greatest mountains in the world had overwhelmed the lowe villages or marshy grounds.[1]

More than 2,000 people drowned as the water drove far inland. You can get a sense of what happened from the contemporary woodcut shown in Figure 11.1. There is also evidence that these monstrous waves, almost certainly tsunamis generated by a sea-floor landslide, washed over islands in the Bristol Channel like Sully Island stripping all the loose material from them, leaving behind only bare solid rock.

Something similar once happened on the coast of Alaska, at a sinister beguiling place named Lituya Bay: *'peut-être le lieu le plus extraordinaire de la terre'* ('perhaps the most extraordinary place in the world'). In such terms did the French naval commander Jean-François de La Pérouse

Figure 11.1 *The effects of giant waves in western England in the January 1607 flood, from a contemporary pamphlet 'A true report of certaine wonderfull overflowings of waters'.*

extol the beauty of Lituya Bay when he saw it first on 13 July 1786.[2] For it is indeed a magnificent sight. With a stage backdrop formed from a sheer cliff above which rise snow-capped mountains, the almost symmetrical bay and its centrally placed island are breathtaking. The bay has succoured many sailors by providing a calm-water haven in this part of the northernmost Pacific Ocean.

So captivated was La Pérouse by his discovery that he became intent on securing Lituya Bay as a French base in Alaska, even to the extent of naming it *Port des Français*. But La Pérouse altered his opinion somewhat when 21 of his crew drowned in the tidal bore as they were taking soundings at the entrance to the bay: an event that led him to erect a monument to them on the island in its centre, which he named *Île du Cénotaph*, Cenotaph Island.

Lituya Bay neither became a French base nor indeed anything other than a temporary haven for mariners, which was fortuitous for its beauty and placidity hid an almost unimaginable terror. The rock walls at the head of Lituya Bay expose the steep face of the Fairweather Fault, occasional movements along which often unleash massive rock collapses, their effects amplified to phenomenal proportions within this closed bay. It is known from both oral histories of the Tlingit people and from later research that movements sufficient to generate large waves in Lituya Bay occurred on four occasions between 1853 and 1958. Having long occupied the region, the Tlingit have a story explaining that a monster living in Lituya Bay periodically grasps the surface of the water as if it were a sheet and shakes it vigorously.[3]

Significant movement along the Fairweather Fault occurred on 8 July 1958. An earth tremor triggered a rockslide which crashed into Lituya Bay. So voluminous was this slide that it displaced a mass of water sufficient to produce a surge of water reaching a mindboggling 524m above the normal water level – skyscraper size. No one measured it at the time but geologists who visited the area a few weeks later calculated this figure by noting the elevations at which trees had been flattened by the passage of this water surge (Figure 11.2). It subsided into a 30m high wave that tore seawards down the length of the bay, washing completely over Cenotaph Island and destroying three fishing boats nearby; remarkably the people on two of them survived.

The contrast in Lituya Bay between its usual condition of tranquillity and occasional sudden terrifying bursts of destruction underline why it is so difficult for us today, just as it was for La Pérouse in 1786, to detect these. Without

Figure 11.2 *Lituya Bay, Alaska, a few months after a 524m high wave ran up the sides of the bay (forest trimlines shown by broken lines) and a 30m wave overran Cenotaph Island.*

almost constant reminders, it is easy to forget such infrequent events. Things were not always thus. Like the Tlingit, many ancient societies established effective early-warning systems of disaster based on often thousands of years of experience, packaged and transmitted across the generations through the regular retelling of oral traditions. Some places have shown themselves to be dangerous and are best avoided.

As for the now-vanished island of Tolamp discussed in Chapter 9, the connection made in ancient stories (derived from eyewitness accounts) between the impacts of large waves and the disappearance of an impacted island is often misleading, a misdiagnosis of what actually happened, albeit a readily understandable one. However high they are, large waves cannot actually destroy islands like Sully or Cenotaph made from hard bedrock. The appearance of this happening is because, as with Tolamp, an earthquake both causes an island to sink suddenly *and* generates tsunami waves, which run across the sinking island as though – to all appearances – washing it away.

But there are other islands, typically low and formed from unbound sands, which can be erased, partly or wholly, as a result of encounters with uncommonly large waves. An example comes from Lisbon in Portugal, which was hit by an earthquake on the morning of 1 November 1755, an unseasonably warm day. At 9.40 a.m., the city literally rocked. Then, as the survivors scrabbled their way to safety, three giant waves came pounding up the River Tagus into the heart of the city.

> On a sudden I heard a general outcry. 'The sea is coming in, we shall all be lost'. Turning my eyes towards the river, which in that place is nearly four miles broad, I could perceive it heaving and swelling in a most unaccountable manner, as no wind was stirring. In an instant, there appeared a vast body of water, rising like a mountain. It came on foaming and roaring, and rushed towards the shore with such impetuosity that, although we all ran for our lives, many were swept away.[4]

The Lisbon tsunamis battered many other coasts in north-west Europe, even reaching across the Atlantic to the Caribbean where they tore across several Antillean islands. Since it occurred so long ago, it is impossible to reconstruct all its effects although, thanks to the existence of maps from both before and after, we do have good evidence from the Algarve coast of Portugal that these monstrous waves shredded a long continuous sand barrier there, instantly destroying many islands while creating more from their remains.[5]

Perhaps the best-documented examples of islands to have disappeared as a result of large-wave impact come from the Gulf of Mexico. Strung out off the coast of the southern United States are lines of low islands. Made from superficial materials – mud, sand and gravel – most of these islands are parts of former river deltas.

Such deltas were more extensive 18,000 years ago when sea level was 120m lower than today. Subsequently, as sea level rose, deltas have become slowly submerged. The lines of low islands we see today off the mouths of large rivers in places like the Gulf of Mexico are what are called barrier islands, the remains of ridges of sand and gravel driven by ocean waves on to delta surfaces when sea level was rising and fortuitously preserved after it stopped, allowing vegetation to gain a tenuous foothold, stabilising the land.

One once-famous chain of barrier islands in this area was the Chandeleurs, occupying the edge of an ancient Mississippi Delta lobe known as the St Bernard that was cut off from the mainland more than 1,000 years ago by a westward shift of the river mouth. The earliest inhabitants of the Chandeleurs grew plump on the fish and seabirds for which the islands became renowned, at least until 1839 when one unflattering account described the Chandeleurs as 'little more than heaps of sand covered with pine forests'.[6] Yet that description shows how much they have changed subsequently, the heaps flattened, almost invisible, the pine forests long gone.

These changes have largely been the result of two processes – sinking and erosion. All who have studied the Chandeleurs are agreed that subsidence has been a significant reason for their disappearance within the past 200 years or so. There is no surprise in this for, as elsewhere (see Chapter 3), the deltaic sediments on which these islands are founded are being compacted as well as sinking, as their great weight slowly reshapes the underlying Earth's crust. But this is not the whole story.

Erosion has been nibbling away at the Chandeleurs ever since they have been part of recorded history, becoming more evident after about 1853. Following a hurricane (tropical cyclone) in 1915, so much land was lost that the islands' inhabitants moved elsewhere. In 1979, Hurricane Frederic scraped past the Chandeleurs, storm waves inundating the islands and creating numerous areas of sediment washout. But amazingly, many of these pock marks later became refilled by the same sediment that had been washed out of them, thanks to ocean-water movements in this part of the Gulf of Mexico.

The increasing frequency and intensity of tropical cyclones in the Caribbean and west Atlantic over the past few decades has threatened the existence of many barrier islands along the shores of the Gulf of Mexico. The end of the Chandeleurs was foreshadowed in 1998 by Hurricane Georges, which sliced more than a hundred channels through the shrivelled islands. Subsequent plans to heal the islands, to restore them to their former state, may have been driven less by ideals of nature conservation and more by the understanding that, were the Chandeleurs to disappear, then an important first line of defence for the south-east Louisiana coast (especially the city of New Orleans) would be lost.

It was all in vain. The Chandeleur Islands were largely erased from the map following Hurricane Katrina in 2005. The islands were reduced to uninhabitable shoals and the iconic lighthouse that for 109 years had warned mariners to avoid this area of shallow shifting sands finally toppled over.

I want now to take you to a place where an island once existed, but where today you can see nothing but submerged coral reef. It is one of the islands that once rose from the broad reef – today dotted with marine sanctuaries and reserves in acknowledgement of its unique biodiversity – that sweeps in a broad brush stroke around the southern side of high volcanic Pohnpei Island in the Federated States of Micronesia (north-west Pacific). The island disappeared before its name – Nahlapenlohd – was ever written down.[7]

Many Pohnpeians know where Nahlapenlohd was, but fewer, mostly elderly custodians of traditional knowledge, can tell you all the gripping stories about its role in Pohnpeian history and how, it is said, it was erased from the geography of this group of islands. The best-known stories about Nahlapenlohd are from the time, about the year 1850, when it was the site of a fierce battle between the rival Pohnpeian chiefdoms of Kitti and Madolenihmw. While fighting over land and resources was not uncommon, the 1850 battle on Nahlapenlohd was especially memorable because it was the first time on Pohnpei that foreign weapons like cannons and muskets, introduced and operated by runaway sailors (known as beachcombers), were used. Among the details in the stories is a report of warriors sheltering behind coconut trees, suggesting the island was then large enough to support a forest of these. Another explains metaphorically that so much blood washed across the shores of Nahlapenlohd during the fray that the island lost its vegetation cover and disappeared in consequence.[8]

Nahlapenlohd is no more, but it is also one of several islands off the Pohnpei coast to have vanished over the past 200 years or so, probably as a result of large-wave impacts. You might say well, these islands have existed so long, why should waves suddenly start to destroy them now? The answer is that for the past 200 years or so, the ocean surface has been rising in almost every part of the world. So, the level at which waves hit the shore as well as their inland reach have both been increasing, something especially marked in this part of the Pacific Ocean where the pace of sea level rise, at least within the last half century, has been several times greater than the global average.[9]

Another of the reef-surface islands here is named Kehpara, centre of a marine sanctuary famed for its populations of purplish black coral. When I visited Kehpara with Gus Kohler in 2014, one of its handful of residents, Ertin Poll, told me over tea about a nearby island named Kepidau en Pehleng, which had been wiped from existence during a storm about 80 years earlier. Although Ertin never saw this island, he had heard stories about it from his father, how it once had supported a permanent population that moved to the Pohnpei mainland after the island became uninhabitable.

In addition to vanished islands like Nahlapenlohd and Kepidau en Pehleng, many other low islands on the Pohnpei reef have been shrinking in recent years, something attributable to the effects of wave erosion amplified by rising sea level. The prospects do not appear very good although some studies show that low reef islands appear to be expanding rather than shrinking, seemingly contrary to what you might expect as a result of sea level rise.[10]

Of course, it is not only human history that is diminished when islands like Nahlapenlohd disappear. There are other land-tethered creatures that are inconvenienced, sometimes majorly, in such instances. In 2018, massive waves generated by Hurricane Walaka removed an island of ecological importance from the geography of Hawaii, leaving endangered monk seals and green turtles without an essential breeding habitat. No one is certain how the situation will unfold, whether this island – East Island in French Frigate Shoals – will eventually regenerate itself or whether it is permanently lost. The answer is critical for the seals and the turtles who may have to try and find somewhere else to have their pups or lay their eggs, but it is equally important for us to understand how such waves can destroy entire islands.

In a final example of island disappearance plausibly due to the impacts of large waves, let us shift to the seas between the north coast of Russia and the North Pole, in a region where research shows a shift from an Arctic climate to an Atlantic one is currently under way.[11] This shift is being tracked by an influx of warmer Atlantic water that is displacing the cooler Arctic ocean water that once buffeted island coastlines here. Over the past decade or so the most noticeable effect of this change has been a loss of floating sea ice, but there has also been a loss of land ice which has revealed a number of mysteries. Imagine an island covered by ice. As with a cake smothered by icing, we have no outward clue as to what the cake itself is actually like, whether it is large or small, round or square, fruity or spongy. If we pick off the icing (rather than cutting slices as convention demands), only then will we learn the answers. And so it is with ice-covered islands – only once the ice has melted can we discover the size and the form and the composition of the island once hidden beneath it.

In the Kara Sea off the north coast of Russia, air temperatures over the past 30 years have been rising faster than the global average – almost 5°C a year compared with a worldwide average of less than half this. For the towns and villages along the continental coasts of the area, where infrastructure development has been driven by the discovery, mostly offshore, of oil and gas, the main concern is that melting of permafrost – the permanently frozen layer within the ground – is leading to land instability that will undermine buildings and the infrastructure of extraction and export.[12] Further north in the Barents Sea lies Franz Josef Land where many ice-covered islands exist alongside a few recently denuded of their ice cover, exposed for the first time in perhaps tens of thousands of years. Russian oceanographers have been astounded at the magnitude and rapidity of the changes they have documented; '… it is quite likely that the outlines of Franz Josef Land actually are far from the ones that are depicted on maps', concludes Aleksandr Kirilov, director of the Russian Arctic National Park.[13]

A few years ago, one of the newly exposed islands in Franz Josef Land was Perlamutrovy Island, but it is now no more. Since no one lives permanently in Franz Josef Land, it is not entirely clear what happened to Perlamutrovy, but it seems likely that for most of its life its ice cover had insulated it, permafrost had bound its unstable constituents together. Jump ahead a few years, its ice cover gone, its permafrost melted through contact with warmer-than-usual ocean water and Perlamutrovy is briefly revealed to be a pile of unconsolidated gravel that was washed away, wiped off the map, by the waves of the first winter storms to encounter it.

In the novel *The Disappearing Island* by Corinne Demas, it is related that on her ninth birthday, a girl named Carrie was given a special present by her grandmother. It was a small box that contained a sand dollar (the skeleton of a burrowing sea urchin) of the kind found on beaches in many parts of the world and a note that read: 'To celebrate your birthday we will voyage out to the disappearing island where I found this sand dollar when I was just your age.' The island referred to in this fictional account is Billingsgate Island, a real island that disappeared half a century ago off Cape Cod in the eastern United States.[14]

Named for its abundance of fish after Billingsgate Fish Market in London, this island had an area of just over 240,000m^2 when Pilgrim Nicholas Snow settled there in the year 1644. At one time in the early nineteenth century, there were more than 30 houses, a school, a try-works (for extracting oil from pilot whales) and a lighthouse on Billingsgate Island. But its days were numbered, a significant blow coming when an 1855 storm sliced the island in half, destroying the original lighthouse and leading another to be built on slightly higher ground. The clearest signs of the continuing erosion of the island were seen during storms. Thomas Payne, lighthouse keeper, wrote in his diary on 22 February 1882 that 'the middle of the Island was flooded five feet of water within fifteen feet of Lighthouse ... the Island lost thirty feet'. So concerned were the authorities that the island might indeed completely disappear that they built a 300m seawall around the lighthouse. But this barely halted the land loss.

By the early twentieth century most residents had abandoned the island, many floating their homes in pieces across to the mainland, a process known as flaking. The only people who remained on Billingsgate Island were the lighthouse keeper and a man guarding its valuable oyster beds. The lighthouse was finally abandoned to the waves in 1915 and, although its bare bones can be seen at low tide today, the island was completely submerged in 1942.

So why did Billingsgate Island disappear? After all, before Pilgrim Snow, the island had been occupied for generations by the Punonakanit people. It was not a recent transient formation settled by people who naively deemed it otherwise. So why in the early twentieth century did it disappear? Again, the answer lies with rising sea level. For the global ocean surface, having been comparatively unchanging for several hundred years, started rising in the early nineteenth century – a process that has continued, even accelerated, since. It is almost certain that this ocean-surface rise was a response to air temperature rise, which has simultaneously affected most parts of the world for the past 200 years or so. Although a rise of the ocean surface does not necessarily increase wave amplitudes, it does mean that the same waves are able to penetrate further inland and damage places once considered beyond their reach. This is almost certainly what happened to Billingsgate – nineteenth-century storms drove waves across parts of the island that might have been barely touched by them centuries earlier.

You might think this the same situation as the Chandeleur Islands, but it is not. For with Billingsgate, there is no clear evidence of land sinking as there is for the Chandeleurs. At Billingsgate it seems likely that waves alone, amplified in their erosional power by rising sea level, caused the island to vanish. A similar situation to that of the reef islands off Pohnpei that disappeared recently, East Island in Hawaii, and probably poor Perlamutrovy, all manifest evidence that particular types of landmass can be wiped off the map by large waves, memories of their existence fading with time until maybe they cross into the realm of myth and fiction ... and stand some chance of being memorialised for their entertainment value.

Some seemingly frail islands that may have been with us longer than you might expect are found in Bangladesh, off the mouth of the vast conjoint delta of the three rivers, the Brahmaputra, the Ganges and the Meghna, which forms the northern coast of the Bay of Bengal. You might think the problems these islands are facing, as with the Chandeleurs and others in delta situations elsewhere, are due to the combined effects of land sinking and sea level rise. In the short term, they inarguably are, but the situation is less straightforward here.

On 26 December 2004, a rapid 8,050km-per-hour slippage of the boundary between the Indian–Australian crustal plate and its Eurasian counterpart in western Indonesia led to the Indian Ocean earthquake. Rather like a zip being opened. Uniquely, it seems, this rapid slip was followed by a slow slip that tripled the associated energy release during the earthquake, raising its magnitude from M9 to M9.3. In just 10 minutes, it released as much energy as is consumed in six months in the United States.[15]

The tsunami waves produced by this phenomenal earthquake fanned out across the Indian Ocean with fatal consequences, although in the Bay of Bengal these were slightly less because the earthquake lifted its floor a little, actually enough to permanently raise global sea level a fraction of a millimetre. And it is such uplift, which occurred during other, comparably massive earthquakes affecting the Bay of Bengal, which means that subsidence cannot be uncritically identified as a significant cause of island disappearance along its apical northernmost coasts.

Most of these islands (known as *char*) formed from sands and gravels washed off the delta surface during high (flood) stages of these giant rivers. Becoming isolated like the Chandeleurs by sideways shifts of the river mouths, *char* have become enduring features of the landscape, their fertile well-watered soils like magnets for rice and jute farmers. But these *char* are dynamic in form, their inhabitants accustomed to their shape-shifting. Typically along those sides where sediment-laden river water drifts slowly past, *char* shorelines may grow outwards. Conversely, along those sides exposed to the ocean, especially to long-fetch waves, *char* shorelines are often cut into and recede. In recent decades, driven by an acceleration in the pace of sea level rise, the long-established balance between accretion and erosion has shifted markedly towards the latter, resulting in some astonishing amounts of land loss in these islands that is understandably concerning their inhabitants. Between 1931 and 1977, Bhola Island, one of the largest in this region, gained 85km^2 through accretion yet lost 376km^2 through erosion.[16] Over the past 40 years or so, the island of Manpura has lost a net 34 km^2 of land, over 9 per cent of its total area.[17] The trend is unmistakable although its causes are more complex than might appear at first sight.

The shapes of islands like Bhola and Manpura will continue to change, but their fate may not be as assured as that of similar islands elsewhere. For there are two positives with Bhola, Manpura and other islands at the head of the Bay of Bengal, two aggradational forces that may give these islands longer lives in a situation where sea level is rising across the world. One is the huge volume of alluvial sediment that passes by them, some inevitably adhering to their shores. And the second is the almost unique case of the sea floor on which these islands are built being periodically lifted up during major earthquakes. None of this, of course, speaks directly to the ongoing habitability of these islands, the inherent difficulties of occupying such dynamic places, the threat of flood and storm surge annually present.

In the next chapter, we look at big solid volcanic islands that destroyed themselves in memorable, often culture-shattering, events. But there are other types of volcanic island, generally smaller and far less solid, that have been washed away by waves. To understand how such islands form, you should appreciate that almost every oceanic island began its life as a tiny volcano on the floor of the ocean. Some such volcanoes then grew so high that they were even able to push their heads above the ocean surface to form islands. The volcanoes, of which islands like those in the Hawaii group are part, are among the highest mountains in the world, far larger and taller than any on the continents.

Not all ocean floor volcanoes become islands. Some lurk menacingly beneath the ocean surface, erupting only occasionally, often in ways barely noticed by those of us on dry land. Take the underwater volcano named Monowai in

the south-west Pacific Ocean, between New Zealand and
Tonga, the summit of which usually lies 120m below the
ocean surface. The evidence that Monowai sometimes
exposes itself is compelling. On 17 October 1977, the crew
of a Royal New Zealand Air Force Orion aircraft
photographed it erupting. It is likely that the surface
disturbance shown in Figure 11.3 was largely a result of gas
bubbles being violently discharged from the upper part of
Monowai, but it could also be that the eruption had built
the volcano summit up to a position much nearer the ocean
surface and that the white circular patch in Figure 11.3B
represents the summit of the volcano itself.

And before we leave Monowai, consider the report
lodged by the crew of the ocean-going yacht *Nutra* about
their experiences on the morning of 11 November 1986 as
they were sailing through the area en route from Rarotonga
to Auckland. Early in the morning, around 7 a.m., the sea
surface around them became agitated. Then at 7.10 a.m.,
the yacht was shaken violently. There was an explosion like
a dynamite blast, followed by two one-second echoes,
perhaps tremors originating on the ocean floor.[18] It may
seem extraordinary that a boat sailing across the deep ocean
should be affected by earthquakes far below, but maybe in
this case the source of the earthquake was much closer to
the surface than the average level of the sea floor here.

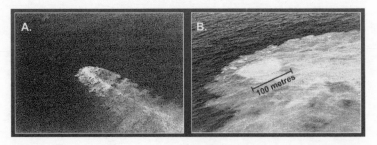

Figure 11.3 *Underwater eruption of Monowai volcano in the south-west
Pacific, October 1977.*

Maybe it was Monowai, grumbling, gastritic, reminding us of its brooding presence.

It seems counterintuitive to suppose that islands can simply appear one day in places where none existed before and then disappear shortly thereafter ... almost as though they had never been there. Like the fabled *abaia anti* of Kiribati or the wandering isles of French Polynesia, described in Chapter 5. Yet this is exactly what happens in many parts of the world where summits of active undersea volcanoes like Monowai grow to within a few hundred metres of the ocean surface, a shallow layer known as the hydroexplosive zone, defined as that within which the weight of overlying ocean water is insufficient to subdue explosive eruptions.

Below the hydroexplosive zone, in the ocean's darker murkier depths, things are different. Here, when magma is extruded from an ocean-floor volcano, it encounters cold ocean water and instantly cools. The weight of the ocean water above it is so great that it prevents an explosive eruption – like holding a handkerchief tightly over our nose to stop a sneeze. In the case of deeply submerged underwater volcanoes, the result is that lava pours out and rolls down their flanks, often forming pillows of lava, characteristic of deep-water volcanic eruptions.

Where the base of the hydroexplosive zone is positioned – and it is not the same everywhere – controls how much of the eruption of an undersea volcano we actually experience. Volcanoes like Kick-'em Jenny, below the surface of the Caribbean Sea near Grenada (of which more near the end of the next chapter) can shoot plumes of ash through the top few hundred metres of ocean and into the air, which is probably around the upper limit of ejaculatory power for most undersea volcanoes.

For aficionados of the pseudoscience writer, Charles Berlitz, his announcement in 1989 that he had discovered an Asia-Pacific counterpart to the Bermuda Triangle – the Dragon's Triangle – must have been welcome. It can be seen a shrewd move by Berlitz as well. For scientists had been uniformly dismissive of his arguments for the Bermuda Triangle, a place where paranormal forces supposedly caused ships and planes to vanish, so the existence of a second triangle might seem to cement the case in the minds of those who really wanted to believe. It is all nonsense.

One of the ships Berlitz claims to have been swallowed up in The Dragon's Triangle is the *Kaiyo Maru No 5*, which was lost along with its crew of 31 people on 24 September 1952. There is very little mystery about why this happened. The ship was investigating a recent eruption of Myojin-Sho (Myojin-reef), an underwater volcano with its top well within the hydroexplosive zone that on this occasion had erupted enough material to produce an island tens of metres high.

Along with several others, Myojin-Sho is a volcano which grew up from the rim of a giant submerged volcanic caldera – the Myojin Knoll caldera.[19] Four hundred and fifty kilometres south of Tokyo, Myojin-Sho is part of the sparsely populated Izu island group, which explains why there is no lengthy record of its activity as there is for most active volcanoes in Japan. The earliest-reported island-building eruption at Myojin-Sho was in 1896, but waves washed away the frail island soon after it formed. The same is true of its subsequent nine appearances and disappearances. Ash and pyroclastic bombs erupted from the underwater volcano formed a pile of material above its summit sufficient to create an island, but this island lasted only as long as the eruption fed it. Once this ended, the waves ate into the loose material and once again wiped Myojin-Sho off the map.

Its last appearance in 1952–53 was memorable for the intensity of volcanic activity, its duration intermittently over one year and the size of the islands it produced on several occasions. At one point, an island 200m long and 10m high formed. The first eruption in the 1952–53 period was reported by the startled crew of a fishing boat, *Myojin-Maru No 11*, for which this volcano became named. Eventually the activity became the source of such concern to the Japanese Maritime Safety Agency that it dispatched its research vessel *Kaiyo-Maru No 5* to investigate. Unfortunately, its arrival coincided with the final massive eruption of Myojin-Sho in August–September 1953. On the evening of 24 September, the *Kaiyo-Maru No 5* passed directly above one of the undersea craters of Myojin-Sho. Gases being erupted were rising rapidly to the ocean surface, decreasing the density of the ocean water to the extent it could not support the weight of a heavy ship, which sank beneath the ocean surface, its crew and passengers lost in an instant.[20]

In 1998–99, the Japan Hydrographic Department conducted a survey of Myojin-Sho using an unmanned survey vessel. It found that the present summit of Myojin-Sho is deeper than it had been in the past, from which it was inferred that the 1952–53 eruption of Myojin-Sho was its terminal one, reducing the height of this submarine monster as well as its potential for activity. No one, however, is placing too much faith in this. On the oceanographic charts of the region, a circle with a radius of 10 nautical miles around Myojin-Sho is designated 'Dangerous volcanic area'.

Half a world away, the inhabitants of the island of Sicily in the Mediterranean Sea have long learnt to live with rumblings and tremblings of the solid earth. For the island is dominated by Etna, one of the world's most incessantly active volcanoes. Etna lies in the east of the island so when, towards the end of June 1831, the people of Sciacca on its south-west coast felt small earthquakes, they naturally assumed these originated from Etna. But this time they did not, as the Sciaccans would shortly discover.

On 8 July, in the ocean some 10km south of Sciacca, water began shooting tens of metres into the air. The ocean surface became agitated, covered with 'reddish scum' containing countless dead fish. A few days later, the ocean surface was so thickly covered by volcanic ash that boats could pass through it only with great difficulty. On 13 July, the people of Sciacca saw 'a column of dark vapour' rising from the sea. After dark, they saw it glowing 'lurid red', and they smelt the stench and heard the sounds of an island being born.

By 18 July, this island had grown up around a crater in which eruptions were seen to be taking place. The island subsequently reached a height of nearly 30m and, after the eruptions ceased, some evidently deemed it safe to land on. The first person ashore was a British naval officer who opportunistically named the island Graham's Island and claimed it, much to the chagrin of its Sicilian neighbours, for the British Crown. No one paid much attention. Later, on 29 September, a French group landed on the island, naming it Julia and leaving us an excellent sketch (Figure 11.4). The French explored the island, in particular the central crater, which was filled with a lake of near-boiling, bubbling reddish water. Around the crater edges were fumaroles from which sulphurous gas hissed menacingly, forming plumes of smoke curling high into the fetid air. The ash composing the island's surface was reportedly hot and difficult to walk across.

The island, which Italian geologists today call Ferdinandea, is no longer visible. By October 1831, all eruptive activity had ceased, the sea had eaten away at its sides until it was no more than 'a hillock of sand and cinders'. Six months later it had completely disappeared.[21] Later, in both 1833 and 1863, eruptions from the submarine volcano that had built Ferdinandea briefly ruffled the ocean

Figure 11.4 *The newly formed island off the south coast of Sicily, as it appeared to Louis-Constant Prévost in September 1831.*

surface, but at no time since the great 1831 eruption has an island reappeared here.

Many of these so-called jack-in-the-box volcanoes periodically appear in the islands of Tonga in the South Pacific. Here in the 1880s, the Reverend Shirley Waldemar Baker was a busy man.[22] Before being deported in 1890, uniquely for being 'prejudicial to the peace and good order of the Western Pacific', he left us an eyewitness account of the 1885 underwater volcanic eruption and subsequent formation of Fonuafo'ou (also known as Falcon Island). Part of Baker's graphic account, which began in the nation's capital, Nuku'alofa, is as follows.

On Tuesday morning everybody's attention was directed to vast clouds of steam and smoke which were arising from the sea in a N.N.W. direction ... others saw a vivid flash of light, and heard a report like thunder ... it was determined that the [ship] *Sandfly* should be sent to ascertain the bearings and extent of the volcano ...

As the *Sandfly* neared the spot the scene was most magnificent, great volumes of steam, of carbonic and sulphurous gas ... being shot forth from many jets out of the sea, in a direct line of over two miles ... to the height of 1,000 feet and more, then expanding themselves in all directions, in clouds of dazzling whiteness, and assuming the most fantastic shapes; sometimes presenting themselves as a mountain of wool, the tips of which were fringed with gold, caused by the rays of the setting sun, then again occasionally forming into a large cauliflower head of snowy whiteness, backed by clouds of intense darkness formed of dust and ashes mixed with watery vapour, which the wind was carrying down for miles on the distant horizon. As the heavier matter kept continually falling, it gradually raised in height the new-made island ...

As the first light of morning appeared ... we found an island had already been formed some three to four miles in length, one in width, and attaining a height of about 40 feet.[23]

Fonuafo'ou – meaning 'new land' – has burst into life at least 11 times since 1781. But we can infer that it has been making a periodic appearance for millennia, probably even being a significant contributor to Pacific people's island-origin myths that may have originated in this part of the Pacific.[24] During most of its known eruptions, Fonuafo'ou formed an island of cinder and ash that, after the eruption ended, was eroded by waves and disappeared.

A final example comes from Solomon Islands, also in the South Pacific, where on 2 April 2007, Marila Timi from Biche Village on Gatokae Island climbed the hills behind the village to weed her taro garden. As she looked out to sea, she saw a plume of dense smoke in the distance rising from the ocean. She knew the oven of the sea god, Kavachi, was alight. Active volcanoes are plentiful in Solomon Islands, some slightly underwater, Kavachi the best known of these. When not in eruption, you can dive and clearly see the outlines of the underwater volcano or, if you are less intrepid, you can just marvel at the tuna-choked waters above its summit. Kavachi is named after a Gatokae sea god and, even when it is erupting, you can sail to within 100m and feel its intense heat and appreciate its synonym, *Rejo te Kavachi* (Kavachi's oven). Fish also feel the effects of Kavachi's activity. When the volcano is erupting, even when there are no signs of this above water, percussion-like noises – whumps – are carried through the ocean as much as 74km away.[25]

Kavachi has a long history of eruption and island-building, certainly far longer than we know.[26] Like all the volcanic islands discussed in this section – Monowai, Myojin-Sho, Ferdinandea and Fonuafo'ou – its origin and activity results from one crustal plate being thrust beneath another. The down-going plate eventually starts melting, the liquid rock forces its way upwards to the surface where it erupts and builds an island.

Most volcanic islands in the world's oceans formed for similar reasons. Some, as will be discussed in the next chapter, have erupted memorably, leading an entire island to vanish, events recalled by science, memory and myth.

CHAPTER TWELVE

Volcanic Islands

In 1507, while a guest of Pope Julius II in Rome, Father Johannes Ruysch drew his map of the world, chiefly remembered today for being the first to incorporate the 1492–1504 descriptions of the Caribbean Islands and northern South America by Christopher Columbus. But it contains something else of interest for this book. In the North Atlantic Ocean, between Iceland and Greenland, it shows an island where there is none today. The accompanying label states that this island was 'totally destroyed by combustion in the year 1456'.[1] There is no scientific evidence that this is true, yet it would be rash to dismiss it out of hand for the Iceland area is indeed uncommonly active in terms of earthquakes and volcanic eruptions.[2] And, given the examples described in this chapter, we can readily believe this unnamed island did once actually exist – before blowing itself to pieces.

And even if we insisted otherwise, there is abundant evidence that Icelandic storytelling was fashioned by its dynamic environmental setting. Imagery of erupting volcanoes, fatally undermining the fabric of society, reminding mortals of the awesome unremitting power of devious and savage gods, permeates ancient Icelandic lore. When Ragnarök (the extinction of the powers) one day occurs, the fire demons will reach the world tree (Yggdrasil) and cause it to burst into flames.

Sól tér sortna,
sökkr fold í mar,
hverfa af himni
heiðar stjörnur;
geisar eimi
við aldrnara,
leikr hár hiti
við himin sjalfan.

The sun turns black,
earth sinks in the sea,
the hot stars down
from heaven are whirled;
Fierce flares the heat
'gainst the life-feeding tree;
till fire leaps high
about heaven itself.[3]

Sometimes, especially if we come from more placid parts of the world, we might wonder at the breadth and depth of imagination we assume to lie behind the creation of such stories. But imagination is unquestionably aided by observation. Seeing, hearing and smelling erupting volcanoes, being scorched by their heat, feeling their pulsations, witnessing their outpourings; all these sensory effects feed our imaginations and help us understand their causes.

Today we might read stories about gods fighting and ripping apart the Earth, causing its insides to come tumbling out, blood red like ours, and think 'how clever' the people who imagined these old stories must have been. But it seems to me unlikely that these people were imaginative in the sense we use the word today. Rather they were scientists, knowledge seekers, struggling just as we do at present to rationalise what they witnessed, to explain why it happened, not least so they could advise their people how best to survive such events should they ever occur again. And this, I suggest, is what lies not just at the heart of many such ancient stories, but also explains the origin of many cultural behaviours in relation to the natural world, ranging from place avoidance to rituals entailing votive offerings at places of violence like volcano summits.

The self-destruction of a volcanic island rarely goes unannounced. Normally, rumblings deep from within the volcano signal the upward movement of bubbling magma. The rumblings warn those who hear and feel them that disaster may be imminent. Montserrat Island in the Lesser Antilles (Caribbean Sea) has never actually blown itself to pieces, but it has an undoubted reputation for rumbling. Until 1995 the local people were largely contemptuous of this, none of the island's volcanoes ever having been known to erupt. But then, on 18 July 1995, came the first recorded eruption of the Soufrière Hills volcano that dominates the island. Contempt turned quickly to apprehension for residents as the capital, Plymouth, was buried in ash. The southeastern half of Montserrat was declared a no-go zone, but it took much longer for all affected residents to be moved out of danger.

Montserrat Island is comparatively small (102km^2) and the Soufrière Hills volcano occupies about two-thirds of this area. While there is no certain prospect of the volcano erupting with such force that it would destroy the entire island, this underlines the fact that many oceanic islands are built largely from a single volcano and that its explosive activity has the potential to affect the entire inhabited land mass.

There are instances where an eruption of an island volcano led to the disappearance of (almost) the entire island, a subject on which this chapter focuses. The first example discussed is famous, not only because it is implicated in the collapse of the Minoan civilisation, but also because it was undoubtedly a major influence on Plato's imagined story of Atlantis. This is the island of Stronghyle, the modern remnants of which are Santorini (Thera) in the eastern Mediterranean.

Like many young volcanoes having a simple geological structure, Stronghyle was a roundish island that attracted people with its rich volcanic soil, while simultaneously unsettling them with periodic displays of awesome power – earthquakes and eruptions. Yet the pull was greater than the push, people wanted to live there and this inspired innovation. It is on Stronghyle that we find the earliest-known examples of earthquake-resistant methods of building, around 3,700 years old. Pieces of wood were inserted in the interstices of stone walls to absorb the effects of shaking during earthquakes and thereby reduce the likelihood of building collapse.[4]

Despite having been home to people for several thousand years, Stronghyle exists no more. Its end, one of the most destructive natural events in recorded history, occurred at some point between 1627 and 1600 BC.[5] The last people to live on Stronghyle were Minoan, part of a civilisation that encompassed Crete, Rhodes and smaller islands in the southern Aegean Sea. The Minoans were a maritime people with an appetite for cross-ocean trade that brought them wealth and allowed them to establish outposts throughout the eastern Mediterranean. The main settlement on Minoan Stronghyle was near modern Akrotiri, which at that time (unlike today, post-eruption) straddled a peninsula, a natural harbour on either side.

Through ancient Akrotiri, there was a trade in copper as well as a hint, suggested by discoveries of smelting crucibles there, that it may have been a place where bronze was manufactured, the copper for this alloy coming from Cyprus, the complementary tin from alluvial deposits in the northern Aegean or even further afield in Hungary or Spain. Ancient Akrotiri was fortuitously located along intersecting trade routes making it an ideal centre for both commerce as well as bronze manufacture. It is a tad ironic that the greatest value of bronze at this time in the eastern Mediterranean was in the manufacture of superior weaponry, yet there is no evidence that the Minoans – at Akrotiri or elsewhere – were especially warlike.

Everything archaeology has revealed about Minoan Akrotiri testifies to its prosperity, at least in the decades prior to its destruction. The town boasted several multi-storey dwellings, their inside walls often painted with frescos,[6] their outer walls faced with masonry, windows overlooking narrow alleys and paved streets beneath which ran a drainage system. Some of the earliest examples of plumbed indoor sit-down lavatories, one at least on an upper floor, were built here as much as 3,700 years ago.

No human remains have been unearthed during the excavations of Minoan Akrotiri, no dead bodies, suggesting that, unlike the situation with the Roman towns of Pompeii and Herculaneum in the shadow of Vesuvius, the island's inhabitants had ample warning of what was about to unfold and had enough time to remove themselves from harm's way.

The first stage of the devastating Stronghyle eruption involved volcanic ash being shot high into the sky, so much that the daytime skies turned preternaturally dark. Then out of the many crater mouths on the island were disgorged huge volumes of fiery rock that mixed with the loose material on the volcano's slopes to create fast-moving pyroclastic flows, some 60m thick, that engulfed (and fortuitously preserved) the abandoned town of Akrotiri. The insides of the former volcanic island were thrown outwards, leaving a massive 60km^3 void below the ground surface into which the remaining fragments of the island slid, creating the largely underwater caldera we see here today (Figure 12.1). Soon after the caldera formed, the ocean rushed in from all sides, a movement

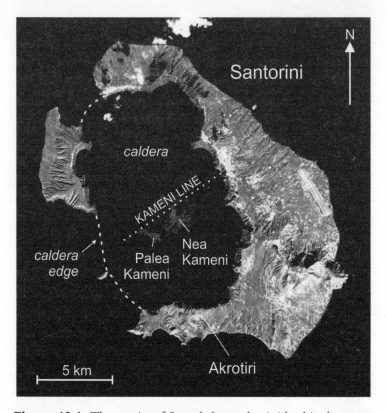

Figure 12.1 *The remains of Stronghyle, a volcanic island in the eastern Mediterranean which blew up some time between 1627 and 1600 BC (around 3,600 years ago). The broken line shows the edge of the caldera into which the island collapsed. Just over 2,000 years ago, new eruptions began along the Kameni Line in the centre of the caldera forming islands that one day may grow to fill the caldera and form a new Stronghyle.*

that generated huge waves which spread out across the eastern Mediterranean, crashing into distant shores with sometimes lasting effects.[7]

Most scholars agree that large waves from the eruption and collapse of Stronghyle fatally wounded the Minoan civilisation through their impacts on Minoan coastal settlements, especially important ones like foundational Knossos in northern Crete. The economic impact of these waves was so great that the Minoans were simply unable to muster sufficient resources in their aftermath to rebuild their civilisation. Their costly attempts to do so left them fatally exposed to Mycenaean aggression and led to the annexation that marks the end of the Minoan era about 1200 BC.

What was left of Stronghyle after the mighty eruption and caldera collapse is pretty much what you see there today: the arcuate-shaped island of Santorini and a number of smaller islands that track the edge of the underwater caldera. The remains of old Stronghyle Island are buried beneath thick piles of pumice and ash and only the occasional discovery of artefacts in a *pozzolana* quarry allows us an insight into the world that existed here before the huge eruption.

Life has returned to these islands, now an iconic tourist destination, but the threat from the volcano has not entirely subsided. For after some 1,500 years of inactivity, the core of the Stronghyle volcano again burst into life. About 198 BC, undersea eruptions began along the Kameni Line (shown on Figure 12.1) leading to the formation of Palea Kameni Island by about AD 47 and its larger neighbour Nea Kameni on 23 May 1707. Although these volcanoes appear only sporadically inactive, they are unlikely to remain so, perhaps one day even starting to build another high stratovolcano like Stronghyle. Given that the underground controls on crustal movement (and therefore volcanic activity) in this area are likely to be the same for the foreseeable future as they were in 1627–1600 BC when Stronghyle blew itself to smithereens, it is likely that there will be one day, far from now, another mighty island-destroying eruption here.

When we think about islands blowing up, we might think of the kinds of cataclysms described by influential writers like Plato and some of his more lightweight successors. In these kinds of (made-up) events, the onset of the disaster is invariably rapid – people have no time to get out of its way – so the death toll is huge, the impacts on culture and society immense and enduring ... and implausible. While we have only a few real examples of this, most such catastrophes appear to have been avoidable by people living in their immediate vicinity, as for the inhabitants of ancient Akrotiri.

So why do we so often marry catastrophe with disaster, link an awesomely large natural event that occurred in the past to a massive human tragedy? Part of the reason lies in the words, spoken and written, of contemporary chroniclers and their successors. To have survived, even if you were far away, a significant event like the destruction of Stronghyle is something remarkable. In ancient Greece, you might conclude that the gods had favoured you. You might be unable to resist the temptation for self-aggrandisement. Indeed, so seduced might you become by your apparent good fortune, you might attempt to boost it by exaggerating your narrative every time you relate it. A volcanic eruption might become an island disappearing, later a 'large' island, later still a continent and eventually 'a chimeric place that takes whatever form its describer wishes to give it'[8] – something that has spawned a vast literature about fictitious Atlantis.[9]

I am sure such scenarios have played out in a variety of cultures at numerous times over the past 10 millennia or more. Although most of these instances are lost to us today, echoes lie in ancient stories about the humanisation of the causes of disaster. In the absence of a science-based understanding of these causes, it was common to attribute them to people's actions. In such interpretations, the disaster

becomes the punishment levelled on delinquent peoples by
the gods they offended or by a person wronged who exacted
a terrible and generally disproportionate revenge on those
responsible, as is shown by the next example.

In isolated communal societies where everyone's survival
is inextricably bound together – a marriage of resource and
effort, supply and demand – individual enterprise is rarely
applauded, never explicitly encouraged. In the early
fifteenth century on the island of Kuwae in the central part
of the islands we now call Vanuatu (south-west Pacific)
there lived a man named Pae,

> … a tall and very strong man. He had long hair and a
> bushy beard and a splendid necklace of shells around his
> neck. On his arms, pigs' tusks and bracelets accentuated
> his muscles … On the entire island of Kuwae, he was
> known to be the best bow shooter. He was talented and
> he never missed his target. Pae was a very proud man.
> While hunting, it was always he who killed the best pig.
> While fishing, the best fish were for him. No one could
> come close to Pae … who was admired but not liked.[10]

Tiring one day of Pae's unrelenting prowess, a group of
young people planned a trick that would make him ashamed
to show his face in public ever again. One dark moonless
night in the common sleeping house (*nakamal*), they
persuaded a woman to lie unknowingly on Pae's mat and
seduce him. After intercourse, the woman felt the raised
scar on Pae's chest and ran away in horror crying 'My son,
my son, I have slept with my son.' Pae cried out, 'My
mother, I slept with my mother. Oh horrible incest,' but in
his head, knowing he had been tricked, a plan for revenge
started forming.

The following day, Pae left his home on Kuwae for the
volcanic island of Ambrym where his uncle lived and

together they climbed to its steaming summit crater. Here were swarms of lizards, vassals to the spirit of this notoriously active volcano, each carrying its magic fire. Frightened by the magnitude of Pae's thirst for revenge, Pae's uncle tried to persuade Pae to select one of the bigger less-harmful lizards to take back to Kuwae, but undeterred Pae selected a small blue-tailed lizard, one 'capable of destroying everything'.

Hidden within a hollowed-out yam, Pae carried this lizard back to Kuwae and buried it near the *nakamal*. Then he asked his brothers to prepare a feast of pigs. The people of Kuwae lampooned him but Pae, stone-faced, stayed silently focused on revenge. After the first pig was slaughtered, Pae took its inflated bladder and hung it from the top of a tall ironwood tree. When the next pig was killed, Pae hung its bladder below the first, and so on until six bladders were hanging in this way, like balloons at different heights on the tree trunk. Once the pigs were eaten, the feast concluded, Pae started climbing the tree. When he popped the first bladder he encountered, the ground began shaking. Even more violently it shook after he popped the second and then the third bladder. The people started panicking, but Pae continued climbing. After he popped the fourth bladder, the land began tilting; the women and children ran into the bush, the elders threw themselves on the ground crying 'Pae, stop! Take pity on us and stop!' But Pae then burst the fifth bladder and the island exploded. He had just enough time to burst the sixth and final bladder, calling out, as a volcano opened at the foot of the tree, 'Your time has come … and my time as well.' Most of the island of Kuwae disappeared into the sea.

You might think that this story, based on an oral history passed on among the peoples of these islands for some 600 years, meant that everyone on Kuwae was killed in this cataclysmic eruption, but this is unlikely to have been the case. For the story of Pae bursting a succession of six pig bladders is better read as a history of precursors, increasingly alarming, of the impending eruption that allowed most people in the area to get out before the climax.

There are other accounts of this phenomenal eruption. One can give slightly less credence to the precise details in that of the missionary Oscar Michelsen, not only because he was writing more than 400 years after it took place, but also because his words, while based on stories he was told, are likely to have been influenced by the visions of hell he routinely employed to persuade the people he adjudged pagan to convert to Christianity. Michelsen wrote:

> ... when all seemed to be peace and safety ... Suddenly there was an alarming subterranean report, accompanied by a violent earthquake. The shock was prolonged into an irregular vibration, and the explosive roar was continued day after day ... Slowly but surely large tracts of land sank into the sea, and other parts of the earth's crust were raised several hundred feet. At three different places, fountains of fire were opened up, and glowing lava sprang into the air to an appalling height.[11]

Kuwae was an island in the peaceful-sounding yet far from bucolic Shepherd Islands in Vanuatu. Reconstructions of the form of the now-disappeared island of Kuwae, achieved by mapping its present underwater caldera, suggest that it may once have been joined to the two modern islands of Epi and Tongoa (Figure 12.2A). The likely eruptive sequence suggests that Kuwae was a tall stratovolcano – like Montserrat or Stronghyle – which exhibited moderate activity from its crestal crater for several months, maybe years, before eruptions increased in violence and the volcano catastrophically exploded, the few pieces left

Figure 12.2 *The island-destroying eruption of Kuwae in 1453 had worldwide effects. A shows the form of the reconstructed island, B shows Sultan Mehmet triumphant outside Constantinople (Istanbul) in 1453, C shows the shallow-water eruption of Karua and D shows the island it formed in 1971.*

behind eventually collapsing into its former magma chamber to form a caldera, today wholly underwater.

Owing largely to its location in a not-so-well-known part of our Earth, far beyond the more comfortably accessible volcanoes of Europe and North America, scientists were slow to apprehend the nature and the global significance of the Kuwae eruption. In fact, some of the French geologists to first study the Kuwae event in the 1990s termed it *l'éruption volcanique oubliée* – 'the forgotten eruption' – to emphasise this point. They were staggered by its apparent magnitude, eventually deeming it 'one of the largest eruptions [on Earth] of the last 10,000 years', something that introduced the name Kuwae to volcanologists' vocabularies the world over.[12]

The terminal Kuwae eruption is estimated to have erupted 30–60 km³ of material, an entire island pulverised and projected skywards with a force 2 million times greater than that of the Hiroshima atom bomb. The Kuwae eruption was also notable for the amount of gas (volatiles) it produced, particularly water vapour, carbon dioxide and sulphur dioxide. When released into the Earth's atmosphere, the latter gas forms sulphuric acid aerosols that often cause short-term (though multi-year) climate cooling. The amount of sulphur dioxide released from Kuwae was so great that this eruption is considered to have been the greatest producer of volatiles anywhere on Earth within the past 700 years. Forty days of unprecedented snowfall occurred in southern China. Across Europe and the Americas, the Kuwae eruption is implicated in a succession of 'years without a summer' that caused harvests to fail and people to starve.[13] Yet the Kuwae eruption also had other effects, ones that may not readily spring to mind.

The capital of Turkey is Istanbul, a modern city overprinted on the ancient one of Constantinople. Itself built on the site of an ancient Greek city, Constantinople was founded by the Christian Emperor Constantine in AD 324, the famed copper-domed church of St Sophia at its centre at once an architectural wonder as well as a symbol of the city's religious importance. For much of its history, Constantinople proved a persistent thorn in the side of Muslim expansionism across the water bodies, shown in Figure 6.2, which link the Mediterranean to the Black Sea. Yet by the mid-fifteenth century, Constantinople had become isolated and could no longer command the timely assistance of Christian leaders in Europe. Understanding this and determined to capture this trophy city, on 15 April 1452 the Sultan Mehmet began preparations to lay siege to Constantinople (Figure 12.2B).

Hostilities commenced almost a year later but, despite the attackers' vast numerical superiority, the defenders held out, upbeat about the prospect of reinforcements arriving. Then on the evening of 25 May 1453, the people in the city saw strange lights dancing on the copper dome of St Sophia; one 'like a large flame of fire issuing forth ... encircled the entire neck of the church for a long time ... [then] took to the sky'. Together with a succession of blood-red sunsets, the devout defenders of Constantinople took this as a sign that their God had abandoned them. The Muslim besiegers, who saw the glow from afar, considered it an omen that the city was doomed – and so it proved to be. By the morning of 29 May 1453, Constantinople had fallen.[14]

A likely explanation for the illusory flames on the St Sophia dome and the blood-red sunsets comes from the Kuwae eruption. It is likely that the lights dancing across the St Sophia dome were St Elmo's Fire, which becomes more common after large volcanic eruptions, even those occurring on the other side of the world, because of the electric fields they generate in the atmosphere. As for the blood-red sunsets, these often occur when there are more aerosols than normal in the atmosphere, something that also often happens after massive volcanic eruptions.

Since it is possible that aerosols from Kuwae took half a year or more to get from Vanuatu to Constantinople, it is possible that this Earth-shattering eruption took place in late AD 1452 although most geologists today regard some time in the first few months of AD 1453 as more likely. Measurement of the relative impact of the Kuwae eruption is possible from contrasting environments, including some on the frozen continent of Antarctica. Since ice accumulates very slowly on the surface of arid Antarctica, faithfully recording the composition of the atmosphere (especially its temperature) at the moment of deposition, scientists have for decades been coring through its ice caps to unravel the nature of climate change over sometimes millions of years. These ice cores also tell us about the timing and relative magnitude of titanic volcanic eruptions such as that of fifteenth-century Kuwae.

The fingerprints of Kuwae are all over Antarctica. For example, in ice cores at now-abandoned Siple Station, close to the Earth's south magnetic pole, an acid spike was found dating from AD 1454–57, marking heightened levels of atmospheric aerosols, almost certainly from Kuwae. In a core from Law Dome, almost 1,400m above sea level and one of the snowiest places in Antarctica, there is evidence for increased deposition of sulphuric acid aerosols from AD 1459–61, the largest signal of a volcanic eruption in these cores over the past seven centuries, also attributable to Kuwae. Significantly, this signal surpassed that of the huge 1816 Tambora (Indonesia) eruption and was roughly six times larger than that of the 1991 eruption of Mount Pinatubo in the Philippines.[15]

The story of Kuwae does not end in the fifteenth century for, although the original island disappeared, the underwater caldera that formed after this happened still occasionally hosts volcanic eruptions, similar to the situation at Myojin-sho (see Chapter 11) and comparable to the new volcanoes stuttering into life in the centre of the Santorini caldera. Sometimes the eruptions from undersea Kuwae produce short-lived islands, although until half a century ago no one was really sure about this.

Late in February 1971, geologist Don Mallick was mapping rock outcrops on the islands close to the underwater Kuwae caldera when he was alerted to its activity. On 22 February he photographed it blowing out ash and hissing steam (Figure 12.2C). The next day, the eruption apparently over, the air cleared and Mallick gaped as he raised his camera to take one of the first photographs of Karua Island, as this evanescent child of Kuwae was christened (Figure 12.2D). The island disappeared soon afterwards, like many of the short-lived islands described at the end of Chapter 11, a victim of the waves eating away its exposed fringes.

In terms of volcanic eruptions and earthquakes, Vanuatu is one of the most active island groups in the world. This is because it marks a place above which the chunk of Earth's

crust we label the Indo-Australian Plate is being thrust
steeply beneath the chunk we call the Pacific Plate;
something that has been happening for tens of millions of
years. A similar mechanism explains the volcanic activity at
Montserrat and Stronghyle/Santorini. In Vanuatu, there
are three parallel chains of islands, the central chain being
that along which all the currently active volcanoes in the
nation lie.

There is some evidence that in recent geological
times, a few hundred millennia ago, the volcanic and
seismic activity in these islands has ratcheted up a notch.
A likely reason is that a line of undersea volcanoes known
as the D'Entrecasteaux Ridge reached the sea floor
boundary between these two plates and has since been
pulled down – under great protest, we might imagine –
along their sub-crustal boundary, locally inflaming the
stresses that cause volcanic eruptions and earthquakes far
beyond what might be considered textbook in such
situations. Similarly, anomalous situations were described
in Chapter 7 for the North Island of New Zealand,
where seamounts can be detected moving slowly beneath
its continental crust, as well as along the Cascade margin
of North America and off the coast of Costa Rica; see
the colour picture section for seamounts here, burrowing
like giant rodents under the continental crust of central
America.

The people who have occupied Vanuatu for at least 3,000
years have developed a unique relationship with their land,
accepting that it regularly changes fundamentally and
violently, sanguine about the associated dangers compared
with most other people on Earth.[16] Yet the hazards in
Vanuatu affect places outside its borders, for which reason
it seems incumbent upon hazard planners elsewhere to
know something of the place. For ignorance may hasten
disaster, as the people of Constantinople unknowingly

demonstrated in 1453, even though they could not have known of the existence of the islands of Vanuatu.

We now shift almost 7,000km west from Vanuatu to the Indonesian islands of Java and Sumatra, which also adjoin a site of active convergence between two sizeable crustal plates. As in the south-west Pacific, so today plate convergence along the Sunda Trench – which lies oceanwards of Java and Sumatra – explains the eruptions and earthquakes that affect this part of Indonesia. Such an earthquake, one of the largest ever recorded, caused the 2004 Indian Ocean tsunami that killed some 230,000 people along almost every part of this ocean's shoreline.

Between long sinuous Java and more bulbous Sumatra lies a number of smaller islands including the volcano named Anak Krakatau, the belligerent 'child of Krakatau'. Its deceased parent, a tall conical stratovolcano named Krakatau (sometimes Krakatoa), is today just a memory, albeit one kept alive in innumerable drawings and narratives. Its 1883 disappearance resounded around the world, literally and figuratively. The sounds of its terminal eruption were heard almost 5,000km away, the ash it produced smothered an area of around 800,000 km^2, and the tsunami resulting from its terminal eruption killed at least 36,000 people.[17]

But what actually happened in 1883? Part of the arcuate chain of volcanic islands that includes Java and Sumatra, the Krakatau volcano had a long history of violent eruption. It began showing signs of activity – rumbling, gurgling, steaming, hissing – three months before its final cataclysmic eruption took place. Eruptions began about five weeks before this, harbingers of what was to come. Then on the morning of 27 August 1883, four enormous explosions occurred, each producing tsunamis as much as 30m high. After the smoke cleared, the survivors, no doubt astonished at their good fortune, were able to see where the island of Krakatau had once been. All that remained was a small piece of what had been the small Rakata cone on the southern flank of Krakatau – and beneath the still-churning sea, a massive, deep debris-filled caldera.

Calderas of this kind have attracted disproportionate attention in recent decades, first because of their hazard potential but also – which may surprise you – because of their economic potential. First, hazards. Terminal, island-destroying eruptions like that of Krakatau in 1883 – and that of Stronghyle and Kuwae somewhat earlier – occur when a chamber filled with liquid rock (magma) below an active volcano abruptly empties, its contents rocketed into the sky and sprayed across surrounding lands. Below the remains of the volcano now lies a void – the empty magma chamber – into which they slip, forming a saucer-shaped landform known as a collapse caldera.

Some time after these island-destroying eruptions, underground magma chambers may start refilling, giving rise to new young volcanoes that may then impudently poke their heads above the caldera, sometimes from its floor as with Palea Kameni and Nea Kameni off Santorini, sometimes from its rim as with Karua. These young volcanoes can eventually grow as large as their deceased ancestors before they too one day blow themselves to pieces – and the cycle starts all over again. In the case of Krakatau, post-caldera eruptions have been concentrated at Anak Krakatau, centrally located within the caldera, that first showed itself at the end of 1927. Erupting occasionally since then, an eruption-associated collapse of the flank of Anak Krakatau in December 2018 generated a tsunami that took the lives of 426 people living on nearby coasts.

Filled with fragments (known as breccias) from the old volcano, collapse calderas are also places where superheated liquids from deep within the Earth's crust sometimes reach close to its surface. Exploiting the fissures and cracks that criss-cross ancient magma chambers, these liquids snake their way upwards until they cool and solidify, often producing deposits of mind-boggling economic value when the hydrothermal minerals they contain are precipitated as veins within caldera breccias.

Venous deposits of this kind are found mostly in above-sea calderas like that of the Tavua volcano in Fiji, from which gold has been mined for some 90 years at Vatukoula, miners following auriferous veins within fault-sliced andesite or basalt dikes, often along narrow underground passages in sometimes intensely hot wet conditions.[18] The situation is quite different when these collapse calderas form below the ocean surface. In such instances, the comparatively rapid cooling that superheated liquids rising upwards through the crust experience when they meet ice-cold ocean water results in precious minerals precipitating out in concentrations as much as 40 times greater than in above-sea calderas. For example, on the sea floor 400km south of Tokyo at the Myojin Knoll caldera, mentioned in the last chapter, there exists a mass of gold and silver 400m in diameter and at least 30m thick.[19]

Long regarded as the emblematic catastrophe, the terminal eruption of Krakatau in 1883 sent shock waves around the world. The livid sunsets that appeared in its aftermath in many places were much commented on at the time and may have inspired the dramatic backdrop to the famously unsettling 1893 painting *The Scream* by Edvard Munch.[20]

The Krakatau eruption, still the largest of modern times, also influenced the way in which people thought about the Earth, its evolution and the place of humans on it.[21]

Industrialising Europe, in the process of its seduction by materialism, was jolted by this awesome reminder of nature's power. Human endeavours appeared almost insignificant in its shadow. Others found evidence in this catastrophe for cycles of human construction and destruction, alternating rises and falls of civilisation they thought to have characterised human history. And in this regard, the 1883 eruption of Krakatau is very important to the central theme of this book – that of land which has disappeared.

Most invented accounts of islands and continents that disappeared include the detail that the act of disappearance was abrupt, cataclysmic: a well-understood tool for sustaining oral stories. I suspect that this detail was mischievously introduced to the land-sinking genre by Plato, but it was rewritten for a modern audience using much of the language used to describe the Krakatau eruption. Consider the end of the supposed lost continent of Mu, claimed to have once stretched across most of the Pacific Ocean (it didn't), described in a 1931 book by James Churchward.

> Cataclysmic earthquakes rent Mu asunder … she became
> a fiery vortex, and the waters of the Pacific rushed in
> making a watery grave for a vast civilisation and sixty
> million people.[22]

Pure flapdoodle, of course. But we can trace self-styled 'Colonel' Churchward's description back to a time a few years after the 1883 Krakatau eruption when Helena Blavatsky, founder of Theosophy, was composing her magnum opus, *The Secret Doctrine*. In it she described a 'huge land' named Rutas allegedly described in (conveniently unspecified) 'Brahminical traditions'. One day, Rutas was abruptly destroyed in a volcanic cataclysm and 'sent to the ocean depths' leaving behind only the islands of Indonesia to mark the place where it once stood. No one has uncovered

Brahminical or any other traditions to support Blavatsky's ludicrous claims about Rutas, but it is almost certain that reports about the Krakatau eruption greatly influenced her thinking at this point in her life as she scratched out her specious legacy in a cramped South London tenement.[23]

So, what happens to island volcanoes after they disappear? After they have blown themselves up? Some, as we have seen, slowly rekindle, sending up shoots from the ashes – places like Nea Kameni adjoining Santorini or Karua which rises from the sunken caldera rim of giant Kuwae. Even the child of giant Krakatau – Anak Krakatau – has already demonstrated a capacity for destruction less than a hundred years after its parent exploded. But when this does not happen, when island volcanoes apparently sink without trace, are they gone forever ... or do they have ways to occasionally remind us of their presence? Some of both.

After the end of the Second World War when the allied navies, hesitant to quickly disband, turned their attention and technology to mapping the deep ocean floor, it became clear that the innumerable undersea mountains they discovered were of two main types. There were conical ones, their peaked tops implying that they were volcanoes that had never broken the ocean surface, and then there were flat-topped ones which were named guyots. It was inferred that the only way that guyots could have acquired a flattened summit was if they had once stuck their heads above the ocean surface, allowing the waves to bevel them or for them to be decapitated by a flank collapse. It is probably a bit more complicated than that for there are tens of thousands of guyots in the world's oceans, but some are likely to represent the remains of volcanic islands that once erupted, thence became submerged and have remained so ever since.

There is some evidence that the submarine volcano named Kick-'em-Jenny (more rambunctious than nearby Kick-'em-Jack) in the southern part of the Lesser Antilles island arc in the Caribbean Sea was formerly a volcanic island that may once have caused its own disappearance through a massive eruption. And while its summit, currently more than a kilometre beneath the sea surface, shows some resemblance to a guyot, this is also pockmarked with young volcanic cones. One of these, having temporarily grown into shallower water, into the hydroexplosive zone, may have been responsible for the only known above-sea eruption at Kick-'em-Jenny in 1939. Since then the summit of Kick-'em-Jenny has lain below this zone, so signs of eruptions have not generally been noticeable at or above the sea surface. Some bursts of activity here have been inferred from the discoloration of the water locally, rather like the situation at Monowai undersea volcano shown in Figure 11.3. But it is likely that we have failed to identify every recent instance in which Kick-'em-Jenny revived.

On 5 August 1944, there was a holiday atmosphere on St George's Pier as many of Grenada's young people were leaving for the Emancipation Holiday weekend on St Vincent Island, 150km away. Two boats were taking them – the *Providence Mark* and the *Island Queen* (Figure 12.3). The more spirited young people were trying to find a place on the *Island Queen*, believing it would be more fun to make the journey on that ship. The two boats left together, a passenger on the *Providence Mark* recalls seeing the lights of the *Island Queen* for the last time at about 8 p.m. as the boats passed Duquesne. The *Providence Mark* reached St Vincent 12 hours later but the *Island Queen* was never seen again. None of the 67 people on board survived and no wreckage has ever been found.

Figure 12.3 *Chicra Salhab, owner of the Island Queen, poses aboard the vessel that vanished in 1944 as it passed over the degassing underwater volcano Kick-'em-Jenny.*

At the time, the Second World War was under way and the initial thought was that the *Island Queen* had sunk following a torpedo attack from an undiscriminating submarine. Some held that a floating mine had become dislodged from the heavily mined harbours on St Lucia or Martinique islands and had floated south into the path of the vessel. The main evidence against these explanations is that no wreckage of the *Island Queen* was ever found, not a piece of wood or a shred of clothing.

A more cogent explanation is that the *Island Queen* passed across the top of Kick-'em-Jenny at a time when this underwater volcano was vigorously emitting gas from its summit. The number of gas bubbles in the seawater may have lowered its density to a point where the weight of the *Island Queen* could no longer be supported and the ship plunged quickly and unceremoniously into a deep watery grave.[24] Just as inferred for the disappearance of the *Kaiyo Maru No 5* at Myojin-Sho, described in Chapter 11.

Islands will disappear in the future for much the same reasons as they have in the past. Aside from recognising the precursors so that people can remove themselves from the danger zone, like they did at Minoan Akrotiri and fifteenth-century Kuwae, there is not much else that can be done. We occupy a geologically dynamic planet where extreme events occasionally occur. Of course, most of us will never get to experience such an event for they are thankfully infrequent.

CHAPTER THIRTEEN

Slipping into the Shadows:
Vanishing Lands

While we sometimes suppress the thought because it often profoundly upsets us, we live in a world where disruptive change is normal, unavoidable and even profound lasting change should occasionally be expected. A farmer may have five continuous years of good annual harvests, but the next may be a disaster. A river may have been flowing through your town for as long as generations of your family remember, but one day it may dry up forever. The beach where you played as a child might be gone when your children's children go to the seashore.

This chapter looks to the future, seeking to understand which lands might disappear underwater in the next hundred years or so. Given that the sea level is currently rising unusually quickly, something likely to continue for the remainder of the twenty-first century and beyond, the most widespread cause of land disappearance is likely to be this – submergence. Life on Earth is not fundamentally threatened by this, but unless current generations make radical adjustments to their expectations, they are likely in the course of their lifetime to receive several sharp and unwelcome reminders about the paramountcy of nature – and the comparative helplessness of humanity.[1]

For coastal dwellers, the most common reminders are likely to involve flooding. The efficiency with which rivers carry water from the land to the sea is premised on the river channel having a seawards slope down which river water can run. Sometimes with lowland rivers this slope is so gentle, almost imperceptible, that water appears sluggish, barely moving. Forgetting for the moment about tidal fluctuations, consider that this movement requires the ocean surface to be lower than the surface of the river water. Yet as sea level has risen over the past few decades so, at the mouths of many rivers the water surface has become higher than the surface of the river water (especially at high tide) so that the slope of the water surface is reversed, forcing river water back across the adjacent land. In a nutshell, rising sea level has increased the frequency, the duration and the extent of lowland flooding around the mouths of most large rivers.

You might consider that this simple idea, linking sea level rise to increasing lowland river-mouth flooding, to be uncontroversial, but it has been resisted in many ways – including the use of abuse and threats, sham science, even stark ignorance – in numerous situations over the past few decades. Behind this resistance, I suggest, lies a fear of the need for what scientists call transformative adaptation, the fundamental (rather than short-term and superficial) response that people living and working in such places will eventually have to make as a response to the continuing rise in sea level. It does not really matter whether this transformation is building massive sea dikes, high walls to prevent the sea encroaching on the land, or whether it is slowly shifting houses and infrastructure from the most vulnerable spots to others less so. Inconvenient as this is, we cannot continue as we have.

Rather than allowing even a whiff of hyperbole to linger after that last paragraph, let me give you a personal example. Thirty years ago, the town of Nadi, the bustling coastal tourist hub in the Fiji Islands, rarely flooded despite being located on the seaward edge of the delta of its eponymous river. But then, around the start of the twenty-first century, it started flooding more often, practically annually. It still does. Nadi residents and many others, locals and tourists, cried out loudly for a solution.

Identifying a solution first required identifying the cause of flooding. There was no shortage of suggestions. 'Poor drainage' was a popular one, the idea being that floodwater in the streets of Nadi could not drain away to the sea because there were not enough drains and many became quickly blocked. The Nadi Town Council spent thousands of dollars on upgrading urban drainage, but this did not stop the floods. Another proposed cause was unspecified 'farming practices' that allegedly had released so much silt into the rivers that their channels had become clogged and could no longer accommodate flood discharges. River dredgers were duly deployed, but this did not stop the floods. Some blamed 'deforestation' of the upper catchment, but this had been deforested decades earlier. Some blamed the removal of boulders from stream beds to reclaim land for tourism development for increased river discharges. Secure in his ivory tower, a Fiji-based academic opined that the Nadi floods were due, as in his native Bangladesh, to seasonal snow melt (in a country that has never recorded snowfall!). The Lord Mayor of Nadi reminded everyone that it was God who had the final say.

I waded into this maelstrom to point out that sea level was rising off Nadi at a rate of some 3–4 mm per year and that in my view this was the principal underlying cause of the increasing incidence of flooding. Unfortunately, this message was not well received. There were unpleasant phone calls and letters, public calls for me to be sacked from my university position (thankfully unheeded) and a torrent of 'well you could be right, but I still think' missives. Some of the most acerbic comments came from coastal developers building and extending areas of reclaimed land in places like Denarau and Naisoso, who were horrified at the negative impact my comments might have on their efforts to attract investors. With time passing, the flooding continuing, all the short-term solutions having failed, you might think it was abundantly gobsmackingly clear that rising sea level was indeed the underlying reason for worsening Nadi flooding and, as a consequence, this was likely to worsen further in the foreseeable future.

Then a crumb of hope was thrown into the mix. As part of their bilateral aid assistance to Fiji, the Japanese government announced it was going to divert the mouth of the Nadi River to a position a few kilometres north of its present outlet in the heart of Nadi Town, thereby removing any risk of future flooding there. I have tried really hard to fathom the reasoning behind this, but I cannot. Suffice to say that the sea level is rising at the same rate all along the shores of Nadi Bay so wherever the river mouth is located it will not make any difference to the flooding of Nadi Town.[2]

Rising sea level does not merely flood the land. It also eats it away. Sandy shorelines gradually adjust their profiles to the average sea surface level. When this level goes up, the sand is removed by waves from the top of the shoreline (where the beach is) and redeposited at its foot, usually underwater. So rising sea level causes shorelines to retreat. And people can directly impact shoreline stability by either removing protective coastal vegetation like mangrove forest or planting more of it. Nature-based solutions of this kind are being increasingly hailed in many places as the best counter to the effects of rising sea level.[3]

Land will also vanish in the future for reasons unrelated to changes in sea level. These include crustal movements, volcanic activity, and even large waves. These are not possibilities that should automatically make us uneasy or panic, but they are topics we all benefit from understanding. For with understanding comes appropriate responses, the best of which are anticipatory rather than reactive.

We often hear that the world faces a profound and unprecedented crisis from climate change. Yet crises of this kind have many precedents in the past, although the pace of this one is unquestionably profound and probably unprecedented. We often read or hear that in the future entire cities will be submerged, coastal lands lost forever, islands will disappear, even entire countries may vanish. Is it true? This is the key question. Will these things really happen?[4] The honest answer is 'we don't know'.

No one can tell exactly what the future holds, neither saffron-cloaked seers in smoke-filled rooms nor informed insightful scientists at their batteries of sophisticated information processors. The future is by definition unknowable, but that does not mean it is unpredictable. What we can predict about the Earth's future comes largely from two processes – analogy and simulation.

We can draw analogies with the past to say what is likely to happen in the future. This is the basis of most hazard warning systems. A volcanic eruption or an earthquake or a tsunami occurred here once and had these particular effects. Should such an event happen here again it is likely to have similar effects, so people should watch for these warning signs, take these precautions. Analogy has been used by people for countless ages. Oral traditions allowed the people of Simeulue Island (Indonesia) to get out of harm's way minutes before the 2004 Indian Ocean tsunami struck it; casualty rates were far less than on nearby coasts for this reason. Similarly, we have learned to recognise the warning signs of impending volcanic eruptions – the likely reason behind the low casualty rates when Stronghyle and Kuwae erupted (see previous chapter) – and in places like Mount St Helens, perhaps the most active volcano in the United States, are prepared to stop people from entering areas likely to become dangerous when these signs are manifest.

Simulation involves making a realistic model of the natural world and then inputting likely future values of particular variables to see what happens. Some of the most compelling climate simulations are the General Circulation

Models (GCMs) operated by more than ten groups of climate scientists in different parts of the world. By entering a range of future values for different variables like greenhouse gas emissions, GCMs produce projections (not 'predictions') of future temperature change and sea level change, labelled with terms such as 'likely' and 'almost certain' as a measure of their probability.

One of the biggest sources of uncertainty in climate projections is how humanity will change the face of our planet over the next few decades, so projections are currently linked to one of four scenarios known as Shared Socio-economic Pathways or SSPs. These include SSP1, the 'Green Road', premised on a future marked by efficient and sustainable use of resources accompanied by effective adaptation, and SSP3, the 'Rocky Road', where we not merely repeat, but amplify, the mistakes of the past around resource exploitation and drag our heels adapting to the unavoidable impacts of climate change.

The most recent assessment report of the Intergovernmental Panel on Climate Change (IPCC), a group established to dispassionately assess recent science about climate change rather than proffer partisan views, states that, depending on SSP, it is likely that by the year 2100 our planet will warm anything between 1.5°C and 4°C (compared with the 1850–1900 average), while the ocean surface is likely to be as much as 82cm higher (compared with the 1986–2005 average). Published in 2014, this assessment report will soon be displaced by another, due in 2022, that will almost certainly adjust the projection for 2100 sea level upwards.[5] Right now, I consider 120cm to be a reasonable upper limit.

The possibilities of such things happening to our world pose huge challenges for us all. Wherever we live, whatever we do. Yet for the purposes of this book, only that around sea level rise is really relevant. To put it clearly in perspective, consider that average global sea level rose 19cm between 1880 and 2012, a period of 132 years, and that it is likely to rise 120cm by 2100. Almost a tenfold acceleration. These two things are discussed next – the last 19cm and the next 120cm.

A sea level rise of 19cm in 132 years may not sound like anything to worry about. It is certainly not for coastal dwellers who live on high ground, perhaps atop cliffs, but for people who occupy low-lying coastal areas it has been a different story. The people of Naicabecabe Village on Moturiki Island in Fiji in the south-west Pacific, have lived there for as long as anyone can remember, but now they wonder how long this can continue. During almost every high tide, the sea wells up inside the village, entering those houses unprotected by raised stoops. The beach has disappeared and land plants close to the coast are all dying or dead. Respectively a result of inundation, shoreline erosion and groundwater salinisation, all attributable to 19cm of sea level rise.

When I was last on Moturiki Island, my research team left from nearby Navuti Village at high tide, and the villagers gathered on the seawall to wave us off, leaving a haunting image of how close their settlement is to the ocean surface (Figure 13.1). Some 15 years earlier, affected by shoreline erosion that was literally eroding the fabric of their community, the people of Navuti built the seawall as a last holdout against the rising water. Twenty years hence it is likely to be overtopped, redundant.

The last 19cm of global sea level rise has been monitored for decades. The advent of satellite altimetry[6] in the early 1990s not only allowed us to measure its rate with unmatched precision, but also for the first time gave us a

Figure 13.1 *August 2001, departing the Fiji village of Navuti on Moturiki Island.*

clear picture of its variability. For sea level is not rising along every part of the world's coast at exactly the same rate. A global average is exactly that.

In some places, the rate of recent sea level rise has been less than average. This has been the case for half a century off much of the western seaboard of the Americas, North and South, probably because of the enduring Pacific Decadal Oscillation that pushes warmer surface waters (which fill more space than cooler ones) from the eastern part of the Pacific to its western parts. Where, as you might expect, the rate of recent sea level rise has been far higher than the global average. In some Western Pacific island groups – in parts of Micronesia and Solomon Islands, respectively north and south of the Equator – so fast has sea level been rising for 50 years or more that it is implicated in the loss of entire islands like Nahlapenlohd, discussed in Chapter 11, affording us an eerie glimpse into the future.[7]

A century from now, when scholars look back and attempt to disentangle the myriad influences on the muddled responses of humans today to their growing awareness of climate change and sea level rise, I suspect that – along with the widespread vacillation and indecision, the vapid defence of self-interest, and the reckless witlessness implicit in denial – the considerable regional variations in the visible effects of climate change will also be identified as key. Why is the sea surface rising three times faster off New Orleans than off Los Angeles, and how did this affect political support for global initiatives to address climate change? Why is the south-east coast of Australia having to absorb the effects of a sea level rising perhaps six times faster than it is off the south-east coasts of India and Sri Lanka? Such unevenness is a burden for coastal planners, a headache for politicians and pure oxygen for those seeking reason to sow doubt in others' minds.

Turning our attention to the next 120cm, bearing in mind the disruption that 19cm has already caused, we start to apprehend the full enormity of the challenge posed by rising sea level to the ways we live. Naicabecabe and Navuti are two among tens of thousands of long-established coastal settlements, from cities to hamlets, across the planet that will likely be uninhabitable in their present condition half a century from now. The rise of sea level will accelerate, more and higher seawalls will inevitably be built, but in the end communities such as these will have to move elsewhere. Seen with detachment, it is obvious that the most sensible thing would be to relocate now, in anticipation of the inevitable. Yet few people think like that, not just because it is difficult to fully relinquish our innate scepticism about knowing the future, not just because of the expense and inconvenience involved in relocation, but also because most coastal dwellers these days live under an illusion of permanency. 'The sea is the sea, the land is the land, that cannot change.'

Well, it will. Shorelines will migrate landwards this century, beaches will disappear and coastal lands become more saline.

The degree to which all this will impact coastal dwellers depends entirely on how they respond. Judging from the past, of which more in the next chapter, many future responses are likely to have only short-term goals, maybe building a range of artificial structures along shorelines or raising dwelling floors above some arbitrarily defined flood zone. Longer-term strategies, the transformative kind, are obviously much better, saving both money and angst.

For instance, along many tropical coasts, it has become clear that mangrove forests are far better investments for stabilising eroding shorelines than artificial structures. Mangrove forests are dense and impenetrable, they bind loose sediments together, they pose a formidable barrier to wave attack along the shorelines they fringe. Under the right conditions they may even be able to migrate inland as the sea level rises.[8] In contrast, artificial structures, particularly the vertical impermeable seawall that is ubiquitous along eroding tropical island coasts, typically create more problems than they solve, often collapsing a couple of years after their gala openings.

Some governments have embraced a degree of long-term planning for future sea level rise. In the UK, where some 2.4 million coastal properties are at risk from floods and an estimated 28,000 may be lost to shoreline erosion by the year 2060, plans have been developed for a staggered withdrawal of people from the most vulnerable parts of densely populated coastal areas.[9] In the conterminous United States, where somewhere between 1.8 and 7.4 million people could be displaced by sea level rise before the twenty-first century ends, there does not appear to be any such nationwide policy in force.[10] And in the Philippines, the 2001 Manila Bay Coastal Strategy, which ambitiously focused on how to move the homes of the 230,000 people in the area most exposed to future sea level rise, has now morphed into a programme concerned less with relocation than with rehabilitation of degraded environments. As with the UK, where there appears to have been little talk about 'staggered withdrawal' for a decade or more, I suspect the retrograde evolution of the Manila Bay Coastal Strategy is a surrender to the overwhelming complexity and cost of relocation reinforced by the economic weight of vested interests.

Global population is also increasing, by a net 81 million people (1.05 per cent) or thereabouts in 2019, so the effects of a nearly tenfold increase in the future rate of sea level rise cannot be viewed as a challenge in isolation. Many of the world's coasts most vulnerable to sea level rise are also densely populated, especially giant low-lying river deltas like those of the Mekong (17.5 million people), the Nile (39 million), and the Ganges-Brahmaputra (108 million).[11] As population densities increase, so too do the demands of people on the natural environment and many once-productive ecosystems have become degraded as a result of human mismanagement forced by population growth. People in many coastal parts of the tropical world will face a profound food crisis around the middle of this century when coral reefs lose their biodiversity (because of prolonged and cripplingly high ocean-surface temperatures and acidifying ocean-surface waters), as well as their keystone role in sustaining nearshore ecosystems on which many communities routinely depend for food.

This situation sounds depressing, but it will prove so only if people sit on their hands and wait for it to happen. Many of the university students I have taught over the past few decades come from islands in the Pacific Ocean where coastal geographies are likely to be radically altered forty years hence because of rising sea level. In my lectures about climate futures they invariably look depressed so I try to cheer them with the thought that 'we are pessimistic only because we can glimpse the future. Five hundred years ago our ancestors could not see the future as we do, so they did not worry as we do today. We should relish the challenge, we should use our ingenuity to find solutions to these problems long before they are upon us.' At this, some of the students, uncomfortable with the responsibility I am hefting on to their shoulders, manage awkward smirks.

But the point is clear. Humanity has been confronted by seemingly insurmountable challenges at various points in history. The Black Death, the Kuwae Eruption, the Störegga Slide, world wars, Covid-19. As a species we have endured, just as we shall endure in a world altered by climate change.

Next, we will look at two examples of what particular places have experienced as a result of the last 19cm along with what they can expect as a result of the next 120cm. We start with the Gold Coast, arguably Australia's leading recreation hub, beloved of surfers and sun-worshippers, where the iconic beaches show signs of disappearing. Then we move to the Florida Keys, the string of around 1,700 low sand islands from Key Largo to Key West off the southwest tip of the United States, the integrity of which is palpably threatened by rising sea level as well as increasingly strong Atlantic hurricanes (tropical cyclones).

Anyone visiting the pulsating heart of Australia's Gold Coast for the first time might not see easily past the iconic beaches, the skyscrapers and its many other material distractions. But as you would expect of a city sprawled across more than 50km of what were once coastal swamps, the Gold Coast is an uncommonly watery place. The barrier beaches shunted into place along the coast by the sea level rise that followed the last ice age are the main reason for the post-war growth of the area. Today, arguably overcrowded and overdeveloped, the City of the Gold Coast and its spectacular beaches are mostly artificially maintained, testimony to the ingenuity of coastal engineers. Pumps control the water level in the ground, gauges monitor water quality; sand slurry is extracted from waterways to maintain their navigability and dumped on eroding coasts; beaches are backed by solid walls, replenished regularly with offshore sand to maintain the environments that each new generation of visitors expects to enjoy here. But with rising sea level and the coastal changes it causes, the maintenance of popular expectation is becoming ever more challenging and expensive.

A turning point in the history of coastal planning on the Gold Coast came in January 1967 when Tropical Cyclone Dinah (followed that year by Barbara, Dulcie, Elaine and Glenda) wreaked havoc across the area, triggering severe coastal erosion. The 1967 event gave a peek into a future for the rapidly urbanising Gold Coast that few had hitherto anticipated. Parts of the Esplanade at Surfers Paradise collapsed while storm waters penetrated far inland, as high as 7m above sea level. Some 8 million m^3 of sand were lost from Gold Coast beaches (Figure 13.2).

In response, local authorities constructed the A-Line, a solid seawall, draped with artificial dune, intended to withstand a one-in-100-year storm event, as the cumulative 1967 events were calculated to be. No buildings were to be constructed seaward of the A-Line. While the A-Line proved successful in protecting parts of the Gold Coast beaches, these still often experience massive changes after storms. In their aftermath, huge quantities of sand are moved back to the eroded beaches, often pumped by dredges from offshore sea-bottom depocentres to the shoreline where it is redistributed by natural processes, or sometimes bulldozers.

Given the centrality of the Gold Coast's beaches to its financial viability, there is naturally concern about whether future storm waves, amplified by higher sea level, might irreversibly reconfigure the tenuous balance of water and land along this iconic part of the eastern Australian

Figure 13.2 *Gold Coast, Australia. Main photo shows Surfers Paradise in 2016. Inset A shows erosion management after the 1967 storms involving the dumping of car bodies. Inset B shows the washed-away Esplanade at Surfers Paradise after the 1967 storms.*

seaboard. Another series of storms such as that experienced in 1967 would probably breach the coastal dune system in several places, something that no amount of subsequent beach renourishment could fix. A way forward is to construct artificial reefs offshore to absorb the energy from large waves before they reach the shoreline.[12]

Coastal resort cities like the Gold Coast are found in almost every part of the world, magnets for those seeking hedonist experiences within frames of sun, sand and sea that contrast markedly with the lives they normally live elsewhere. The challenges faced by the Gold Coast are not therefore unique. Places like Benidorm (Spain), Cancún (Mexico), Cannes (France), and Dubai (United Arab Emirates) are similarly challenged, as are a number in the United States including Atlantic City (New Jersey) and Myrtle Beach (South Carolina). Sustaining coastal-focused tourism here is going to become increasingly complex and expensive over the next few decades and it may be that some local authorities will eventually decide that the cost outweighs the benefits. Nature will quietly be declared victorious as alternatives to coastal-focused activities are up-played to try and preserve the reputations of these destinations as desirable places to visit.

There is a memorable exchange in the 1948 movie *Key Largo*, the backdrop to which is a hurricane smashing into the Florida Keys. Ralphie, one of the out-of-town villains, asks Curly, a local hoodlum, as the wind is building up, 'What all happens in a hurricane?' Curly replies, 'The wind blows so hard the ocean gets up on its hind legs and walks right across the land.' Which is an apt analogy, especially if you are living on a slip of land that rises no more than a couple of metres above the ocean surface.

Popular characterisations of the Florida Keys archipelago as exceptionally vulnerable to hurricanes and rising sea level owe much to the problems being experienced on Key Biscayne, the closest of the main islands in the group to Miami, most of the rest of the islands being structurally less vulnerable. The larger of these islands are all formed to some degree from limestone, much of it dead coral reef exposed by a small net fall of sea level within the past 120,000 years. This limestone typically rises a few metres above present sea level and has acted as a focus for the accumulation of sand and gravel carried onshore by waves, explaining why most of the Keys have not to date been summarily washed away by hurricanes. This limestone is also very hard, great for house foundations, which explains why the larger Keys it underlies are often densely populated – and real estate disproportionately expensive to purchase.

The last 19cm of sea level rise have seen both the incidence and magnitude of flooding of the Florida Keys increase, especially during hurricanes. What local residents call 'nuisance floods', which typically occur when the moon is full, have also become more common, prompting local authorities and property owners to invest in ways to reduce their impacts. These include fitting one-way valves to storm drains to prevent ocean water rising up them at such times … and of course raising houses so that they are no longer flooded regularly. But both of these are short-term solutions, effective perhaps for a decade or so, designed only to postpone the inevitable.

Within the remaining decades of the twenty-first century, a sea level rise of 120cm will see at least 74 per cent (119km^2) of the Upper Florida Keys submerged. An estimated 15,933 people will have to move house. Property valued at more than $10 billion will be lost.[13] In the case of places like the Florida Keys, there is no real way to sugar coat this projection. The region is a good example of how relocation will be forced on people occupying such places the world over. No one wants to be told they must move, but anticipating that this will eventually be necessary and doing something about it before it becomes imminent is an appropriately sanguine response.

In trying to capture the range of causes of land submergence over the next century, there is a danger of becoming overly focused on the most easily identifiable. Those that are continuous, readily measurable and comparatively slow. But there is any number of shorter-term changes, often abrupt, of the kinds discussed in previous chapters that might come into play. Since these are far more difficult to predict, their importance to long-term future planning is sometimes downplayed. They fall under the 'acts of God' that feature in the law and in small print in insurance documents. Acts of God are events for which no one can sensibly be held liable, those that could never have been foreseen. But as our understanding of the natural world has grown, foresight is also improving, something with legal ramifications of which people in positions of responsibility are starting to take note.[14]

Compared with human understanding of slower continuous changes (like land sinking and sea level rise), our understanding of abrupt and infrequent environmental changes is far less complete. To illustrate this, three examples are given below. The first is that of an unstable island in the south-west Pacific, a flank collapse of which might send shock waves across this ocean; the second is of a glacier in the Antarctic; the third of an unstable delta front along the Caribbean coast of South America. These examples are more or less random. There are inevitably gaps in scientific knowledge about any future catastrophe-causing situations.

Indigenous myths from people occupying oceanic islands sometimes picture them balanced precariously on rock pedestals that are occasionally shaken, and off which the islands may even sometimes be toppled. From Fatu Huku in the Marquesas Islands in French Polynesia, the story of the island's shark guardian lashing the island's undersea pedestal until it smashed was discussed in Chapter 10.[15] In similar fashion, off the eastern end of the island of Timor, there once existed an island named Luondona-Wietrili that was broken up by the wrath of a large sailfish leaving behind only the smaller island named Luang.[16]

The island of Tanna in Vanuatu in the south-west Pacific is the site of one of the most continuously active volcanoes

(Yasur) in this uncommonly active group of volcanic islands; Yasur has been in unceasing eruption since at least the visit of Captain Cook in 1774. Tanna is also an island that experiences frequent abrupt land uplifts. Missionaries at Port Resolution on Tanna in 1878 wrote about a strong earthquake that produced a 12m-high tsunami and caused a landslide large enough to block the harbour entrance. Then, just a few minutes later, the western side of the bay was lifted upwards a staggering 6m. Another earthquake a month later pushed the land up a further 4m. Today you can amble along the path at the top of this 10m-high terrace and see corals which lived in 1878 before being killed in an instant.

What is especially interesting is that these earthquakes were apparently not felt by people 20km or so away.[17] At first, this seems unbelievable. Shakes of this magnitude went unnoticed by people so close? But neither did they report experiencing the tsunami, so what could be the explanation? Geologists have focused on this detail to argue that the land uplift that Tanna has been experiencing over the past few hundred years is due to 'magma inflation' in the vicinity of Yasur, rather than the more common upward movements associated with plate convergence in such places.

Magma inflation refers to the localised expansion of the Earth's crust as a result of more and more magma (liquid rock) being pushed through it towards its surface. Beneath active volcanoes like Yasur there is a chamber within which magma from deeper down inside the crust accumulates. When that magma chamber is full, either the magma needs to escape (typically through eruption) or the chamber needs to expand (by land uplift), which is essentially what is happening here and in other active volcanic zones like the Campi Flegrei west of Naples (Italy), mentioned in Chapter 4. More than 300 years of continuous eruption at Yasur have clearly been insufficient to prevent magma inflation – more magma is being supplied to the underground magma chamber than is being erupted out of it. Somewhat worryingly, this situation may not continue.

One clue that this may be the case is the distribution of the coral reefs lying off the coast of Tanna Island. Fringing 80 per cent of the island's coast, these are just what every cold-climate visitor expects to find off the shore of a tropical island – wonderlands of richly diverse multicoloured reefs just beneath the water surface extending a couple of hundred metres out from the beach. But on Tanna, the other 20 per cent where there is no fringing coral reef comprises the coast around the foot of Yasur including Port Resolution. The absence of coral reefs around the entrance to Port Resolution, which led Captain Cook in 1774 to adjudge it an excellent harbour, has altogether more sinister overtones for marine geologists. This is because the absence of a fringing coral reef along a tropical island coast is often a sign that the land has either gone up or dropped down recently, taking the coral reef with it. In the case of this part of Tanna, it is the former. Since 1878, because of magma inflation, the land around Yasur has risen so fast here that there has been no time for waves to cut a rock base on which new reef might start growing.

But there is a possibility of something altogether larger occurring here. Tanna Island is bean shaped. Not what you might consider the shape of a classic volcanic island. In fact, it looks like two-thirds of the original island has slipped away to the north-east, rather like the situation in Hawaii (Figure 10.2). This is not something that would have produced just a localised tsunami – it would have been a far-reaching pan-Pacific event – and the key question is 'does the recent history of uplift and volcanic activity at Yasur presage another such event?'

Given that most of the recent uplift and volcanism on Tanna Island has focused in its east around Yasur and Port Resolution, a fault-bounded area named the Yenkahe Resurgent Block, there is a possibility that future rapid uplift (caused by magma inflation) might trigger landslides even larger than the harbour-blocking one in 1878. These landslides might themselves create new lines of weaknesses along which magma could escape the brimful magma chamber below the surface. These are all possibilities, no more. We should not be tempted to overstate the case or to confuse concerns about the possible magnitude of such an event with its imminence, about which we know far less.[18] But there could one day be a catastrophe here involving tsunami waves crossing the entire Pacific and washing over many of its low-lying fringes.

Fifty years ago or more, the ice-encased continent of Antarctica was different to the way it is now. In the past, most parts of the Antarctic coastline were buttressed by shelves of floating ice, but today some of these ice shelves have thinned, no longer able to hold in place the huge masses of ice on the continent as they once did. In several places, particularly at the mouths of glaciers – the 'rivers of ice' flowing out from ice sheets – this has led to a situation in which ice is now sliding off the land and into the ocean, where it melts and causes the sea level to rise.

If all the water currently locked up in ice sheets covering the land were to empty into the ocean, its average level would rise some 65m. History suggests this is unlikely to happen in the foreseeable future. The giant East Antarctic ice sheet in which about 90 per cent of this water is locked has been in existence for at least 35 million years.[19] And it has withstood periods of warmer temperatures than those currently being experienced or indeed anticipated. But its smaller offsider, the not-so-vast-yet-still-sizeable West Antarctic ice sheet, is a somewhat different beast.

The main difference is the level of its rock foundation. The East Antarctic ice sheet sits atop a high continent, mountains poke through in a few places. But the West Antarctic ice sheet rests on top of land that is mostly below present sea level and is held in place by its fringing ice shelves, which is why their thinning is a cause for concern here. There is a possible analogy.

Over the past 2 million years or so, the temperature at the Earth's surface has oscillated between cooler periods (known as ice ages or glacials) and warmer periods (known as inter-glacials) such as that in which we live today. As explained in Chapter 8, the ocean surface is higher during inter-glacials than during ice ages when some ocean water had been converted to land ice. The last inter-glacial before the present one peaked about 120,000 years ago and there is evidence showing the ocean surface at this time reached 6m or so higher than it is today. For decades, scientists have pondered why, where the extra ocean water could have come from. The most compelling answer is that it came from melting of the entire West Antarctic ice sheet ... for the volume of ice this contains equates to exactly the amount of water needed to raise global sea level 6m.

If the West Antarctic ice sheet melted once, it might do so again. A future scenario might unfold like this. A loss of buttressing ice shelves allows the ocean to penetrate beneath the West Antarctic ice sheet, causing it to become buoyant – to start floating. Warmed from below, the ice above starts melting and sea level rises.

When we study the distant past, it is easy to exaggerate the pace of environmental changes. Things often appear to have occurred faster than they really did. To the first scientists to detect evidence for the possible melting of the West Antarctic ice sheet during the last inter-glacial period, it seemed that this event had occurred quickly – 'catastrophic' and 'collapse' were terms commonly bandied about. Today, thanks to the acquisition of more fine-grained data about the past, scientists have pulled back somewhat from using such terms. Which does not of course mean the future threat miraculously disappears.

One place along the edge of the West Antarctic ice sheet where ice is currently being lost at an alarming rate is the Pine Island Glacier. Together with neighbouring Thwaites Glacier, the Pine Island Glacier drains one-fifth of the entire West Antarctic ice sheet and is currently responsible for about 10 per cent of the observed rise in global sea level. The reason for the rapid melting of Pine Island Glacier is its 'ungrounding' – the penetration by the ocean into the gap between the ice sheet and its submerged rock foundations that is leading to their separation. Had scientists not looked harder, the ocean warming of the glacier's underbelly might be regarded as the principal reason for this separation, but it is now clear that Pine Island is not typical. For beneath the glacier, there is a volcano, the heat from which is the major cause of the melting of Pine Island Glacier.[20]

The existence of a volcano beneath a conspicuously melting glacier might seem to acquit climate change as a cause of ice loss in the West Antarctic, but it does no such thing. It merely emphasises the dangers of compartmentalisation, putting things into convenient boxes for the purpose of studying them. A volcano may be contributing to the loss of Pine Island Glacier ice, but ice shelves are thinning all along the coast of West Antarctica, perhaps in exactly the same way as happened 120,000 years ago before its ice cap began melting and sea level rose 6m.

A decade or so ago a flurry of global attention was given to the Cumbre Vieja volcano on steep-sided La Palma, one of the picturesque islands in the Canary group of the eastern

Atlantic Ocean. Signs were detected that a collapse of the western flank of this volcano might be imminent. It was thought such an event might trigger a tsunami that would run up and across the coastal fringes of the Atlantic, including the eastern seaboard of North America, with devastating consequences. Both the imminence and size of this possible event have been downplayed in subsequent studies, although the threat from such events is not trivial, as examples of catastrophic flank collapses of such islands described in Chapter 10 show.[21]

If timing – *when* might something happen – is a source of uncertainty when looking into the future, so too is *where*. For you might think that we pretty much know what is going on in every corner of the world today, especially whether or not a major earth-shaking event in a particular place is imminent or not. That is true to some extent, but it is also something of a false conceit, more peculiar to those who rarely venture from their familiar part of the world than those who purposely seek adventure beyond it. Earth scientists have steadily been chipping away at the mysteries in less well-known parts of the Earth for centuries, but the world is vast, Earth scientists few, their communications often parochial and spatially biased, so inevitably gaps remain.

For me, South America is a closed book. I know the outlines of the continent, names of places, smatterings of its history but, having never been there, it is more like a silhouette to me than a detailed portrait. Yet if I was in the business of risk management somewhere in the Caribbean, I would certainly want to learn more about the Magdalena Fan.

Rising in the Andes, the Magdalena River follows fault-lined valleys through the heart of Colombia reaching the Caribbean Sea near Barranquilla. The Magdalena carries a huge volume of sediment that has infilled the river channels near its mouth and over millions of years created a massive delta.[22] The Magdalena's sediment load has remained consistently high for most of its life because the Andes Mountains from which most of this comes are rising, pushed upwards by the relentless shoving of the Pacific crustal plate beneath its South American counterpart. In other parts of the world, rivers eventually denude mountains, lowering them in height and thereby reducing their capacity for producing sediment. But in the Andes, every time the river starts to denude the montane landscape and lower its surface, up it rises anew, presenting a rejuvenated terrain ripe for denudation once again. Like a kind of geomorphological Sisyphus, the task of the Magdalena is unceasing, almost the only evidence of the longevity of its working life lying in the massive piles of delta muds comprising its 1,700km^2 delta and its undersea extension, the Magdalena Fan.

Research shows that the Magdalena Fan has periodically collapsed, the displacements 'comparable in scale to the largest known landslides on Earth'.[23] The existence of a layer of gas hydrates exactly where the slip planes are located leave little doubt as to their culpability in this instance. Without speculating about *when* another collapse might occur, it is possible to predict its likely effects. As happened with the Störegga Slide, described in Chapter 10, an undersea collapse of the Magdalena Fan would create tsunami waves that would wash back across the land, reaching perhaps 20m in height at Barranquilla. They would also fan out across the Caribbean, a wave around 6m high making landfall at Kingston in Jamaica and Santo Domingo in the Dominican Republic.

At various scales in various places, the abrupt collapse of half of Tanna Island, a major ice surge from the West Antarctic, or a mega-failure of the Magdalena Fan, would cause an unusually rapid rise in the ocean level that would create problems for the current generation of coastal dwellers, urban or rural, almost everywhere. Yet all of these things that might happen in the future have happened before. Many have happened within the 200,000 years that we, modern humans, have roamed the Earth. So, considered dispassionately, you might wonder why we as a species seem to favour living along the coast when it is clearly far more dangerous a place to reside than many others.

Early in 2019, I was part of a research team that stayed for ten days with the hospitable people of Vabea and Waisomo villages on the Fiji island of Ono in the Kadavu group. One evening, at the back of the beach, I was sitting with one of the Vabea elders on an upturned boat watching the sun sinking behind distant islands, bathing us in its orangey glow. We were talking about the perils of coastal living – all six villages on Ono Island are coastal – and especially the effects of rising sea level, yearly eating slowly yet perceptibly away at the shoreline, allowing waves to reach places they never reached before. My friend told me that 200 years ago, no one on Ono lived on its coast; everyone lived in the hills. Mulling this over, soothed by the sounds of the waves lapping gently against the sand mixed with children's distant laughter we sat in silence for a while. Then my friend abruptly broke the mood with a great 'HA!' and I gave him a quizzical look. He explained:

> My grandfather told me that his grandfather was forever saying it was a mistake for our people to leave the hills and move to the coast. In the hills we were safe from the waves, our land was not being eaten from under us like it is today. But we had no choice – you people, you Europeans, you came along and forced us all to move to the water's edge. The old ones, they knew it was dangerous and they told us not to go. But we did, here we are and *dina saraga* – too true – we now discover it is a dangerous place! We need to listen to the past.[24]

Where they have a choice, many people today might express a preference for living along the coast. You might even deem this a long-standing human trait, but it is not. In many places, many cultures, it is clear that until recently coasts were liminal places, exposed, avoided by most people.[25] Today, in the words of John Gillis, 'we are in the midst of a cultural reorientation of vast significance' where within the past century most of the world's landmasses have been 'hollowed out', their human occupants racing to snatch up and settle land on their fringes.[26] The question of what caused this is beyond the scope of this book, but the burning question is whether in fact it might be time for a divorce. Should we wrench ourselves away from coastal living, unhand the seductive sirens who lured hence to these dangerous places?[27]

In some places, divorce will be messy, in others uncontested. One of the latter situations is the coast of Chesapeake Bay in the eastern United States where, as we saw at the start of Chapter 3, the combined effects of rising sea level and land sinking has already led to people abandoning its coasts, as well as entire islands. Marriage counselling is not working. The marshes fringing the Chesapeake coast show no signs of migrating landwards as sea level rises. The idea of armouring the lowest coastal parts of the bay with seawalls to prevent their submergence would actually increase flooding in the upper parts of the bay, affecting the cities of Annapolis and Baltimore among others.[28]

But humans are almost limitlessly ingenious. And as we shall see in the next chapter, they have deployed that ingenuity in a variety of contexts to delay or prevent the drowning of the land, to frustrate nature.

CHAPTER FOURTEEN

Out of the Shadows: Resisting Land Submergence

Looking across the sweep of history, as humans today are uniquely privileged to do, we have seen how our ancestors on innumerable occasions in many parts of the world were displaced from the coastal lands they occupied as these became submerged. Undoubtedly, the most common response was for people to shrug their shoulders in resignation and move somewhere else. Such a response – relocation in modern parlance – was easier in the past, especially the distant past, when humans were more nomadic, their ties to place unframed by law, less tightly constrained by investment and entitlement than is the case today. It was when nomadism was replaced by sedentism that unrealistic expectations about the environmental stability of the world's coasts were born … and have been causing problems for us ever since.

What changed was that people invested time and energy in constructing the trappings of civilisation in one place. Whether these were fences or freeways, temples or tunnels, they made the society that benefited from them instantly more vulnerable to external impacts like rising sea level. A sitting duck, if you like, rather than a moving target. Resisting these impacts and preserving their investments led societies to invent ingenious ways of minimising extrinsic impacts. Two of these ways have long pedigrees. They were employed across the world in the distant past for the same reasons as they are today. They are land reclamation and artificial-island creation.

The rise of sea level across low-lying coastal lands naturally inconveniences its human occupants, especially if population densities are high or much has been invested in the places threatened by drowning. Ways of removing the sea and sustaining the utility and area of the drowned land, even extending it, is what is meant by land reclamation. A convincing case can be made for land reclamation lying at the heart of the economic success of countries like Italy, the Netherlands and Japan over the past 150 years or so, while it plays a growing economic role in many other countries like China and Nigeria. People in England today, driving from Cambridge to Lincoln for instance, might be surprised to learn that they are crossing the Fens, a great marsh, the draining of which began in Roman times and was completed only in the eighteenth century.

Lauded for his resistance to the eleventh-century Norman invasion of England, a great British hero (conceivably Danish) was Hereward the Wake (meaning the 'wary') whose defensive base was the Isle of Ely in the English Fenland. Named for the abundance of eels found in the waterways that once surrounded it, making it an ideal refuge, Ely became an island several thousand years ago as the ocean surface rose across The Wash and penetrated the low delta of the rivers Nene and Great Ouse, saturated the clays comprising it and creating a massive area of marsh known as the Fens. The few hills in the former delta – like Ely – became marsh-surrounded islands, later prized for defence, but also as religious centres, the medieval fear of sectarian persecution lessened by the difficulties of accessing such islands.

While much of the English Fens has now been drained, converting marshlands into flatlands dominated by arable farming, the process of retrieving the region from the sea was not always straightforward, nor indeed fully supported by local residents. In the seventeenth century, especially during the reign of King Charles I, the draining of the Fens was portrayed as key to the national interest, reflecting 'the needs and methods of a burgeoning early modern state'. Drainage, it was said, would allow the agricultural potential of this fertile area to be fully realised and would convert supposedly backward fen dwellers into civilised hardworking citizens. After signs of local resistance to this interventionist agenda, the draining of the Fens became a pet project of Charles who undertook to 'interpose our regal power and prerogative' whenever the drainers needed this.

The people of the Fens did not oppose their drainage as such, but were concerned about the implicit loss of their land rights, for once land was drained it was declared 'improved', liable to 'enclosure' and invariably sold to 'investors' with no compensation for those who formerly used it. People like Sir Cornelius Vermuyden, a Dutchman lured to England to help direct the drainage of the Fens, were compensated for their efforts with title to large tracts of drained fenland. Naturally Fenlanders resisted. For example, in the year 1699, over 1,000 of them assembled 'under colour and pretence of Foot Ball Playing' and smashed pumps, embankments, houses and barns in the Deeping Fen causing £100,000 worth of damage. While this was not an isolated incident, the power of the state eventually won through although many a modern Fenlander probably enjoys a wry smirk at recent initiatives to restore parts of the Fens to their natural condition, allowing in the ocean once again to recreate the area's pre-seventeenth-century ecology.[1]

The situation was similar elsewhere. An eighteenth-century traveller in the North German Plains would have encountered a landscape quite different from today's:

> Dark and waterlogged, filled with snaking channels half-hidden by overhanging lianas and navigable only in a flat-bottomed boat, these dwelling places of mosquitoes, frogs, fish, wild boar and wolves would not only have looked but smelled quite different from the open landscapes of windmills and manicured fields familiar to twentieth-century Germans.[2]

While it is fair to say that memories of the environmental imperatives for land reclamation have been long forgotten by most residents of places like the English Fens and the North German Plains, the situation is different where reclamation schemes have proved less sustainable.

An example comes from the Zhujiang (Pearl River) Delta in southern China, the fertile soils of which have long attracted farmers from its hinterland. Four thousand years ago, the ocean surface in this part of the world rose a couple of metres higher than it had been a few hundred years earlier and much of the Zhujiang Delta went underwater. Many resident farmers were undeterred, it seems, shifting to hilltops fortuitously scattered throughout the delta and cultivating their peripheries, water lapping at the fringes of their fields. But the heyday of Zhujiang agriculture came 2,000 years later, during the Qin and Han dynasties (221 BC to AD 220), when a rapid fall in sea level exposed a vast expanse of land. People poured in from central China bringing with them cutting-edge agricultural techniques, particularly for rice cultivation, that saw the prosperity of the delta boom.

But within this simple framework, the Zhujiang Delta has seen many changes, some attributable to comparatively minor rises and falls of sea level, that elicited innovations both for protecting land threatened with inundation and for reclaiming recently drowned land. Compounding the problem, as in all deltas, was the effect of river sediment inputs, something that also varies over time, both as a result of rainfall changes and as a consequence of upstream land-use changes like forest clearance.

The earliest-known dikes in the Zhujiang Delta were built about AD 850. These dikes were ridges of earth, armoured with rocks where available, that kept unwanted water off farmers' fields. Once the success of this technological innovation became widely known, it was employed more frequently across the Delta to extend the area for agriculture, especially rice. Networks of dike-enclosed fields sprung up all over the delta, double cropping of rice was enabled, and Guangdong rice became a popular staple throughout southern China and adjoining lands. Today huge tracts of the Zhujiang Delta are covered with networks of dikes that simultaneously support fish farming, mulberry cultivation and silkworm rearing.[3]

A comparable situation evolved on the east coast of Canada, especially in the low south-east part of coastal New Brunswick bordering the Bay of Fundy, renowned as the place with the greatest tidal range in the world. Twice a day at the head of the bay, the ocean surface rises and falls an astonishing 16m. In the first decades of the seventeenth century when French colonists arrived in this area, they traded and maintained cordial relations with the Mi'kmaq who had occupied it for millennia. Realising the agricultural potential of coastal marshes, the Acadians, as the settlers became known, began draining these through the construction of dikes known as *aboiteaux*. Three hundred years later, Acadians still occupy the region though, owing to a series of eighteenth-century deportations, their presence is less visible than it once was. Which is why the first-ever Acadian-language movie, the 1955 *Les Aboiteaux*, was so significant. It includes a dramatic scene in which everyone from the local Acadian community, aware that

one of the *aboiteaux* keeping the ocean off their fields is
failing, rushes out with shovels to rebuild it. The evening
light gives way to darkness, but the frenzied pace of the
shovelling hardly lessens. The issue is one of survival, an
apt metaphor for the survival of Acadian culture which is
famed for innovative dike–building (Figure 14.1).

Acadian *aboiteaux* were distinct because they incorporated
sluices containing a hinged gate (*clapet*) that would remain
closed at high tide, preventing the ocean from flooding the
land, but would open when the tide was falling, allowing
excess water from the land to drain into the sea. This
ingenious system allowed the coastal Acadians to drain and
desalinise marshes, exposing deep fertile soils that were
then planted with corn, hay and flax, cherry, pear and apple
orchards, and used to graze cattle.[4]

Another place where history readily morphs into
metaphor is the city of Mumbai (Bombay) on the west

Figure 14.1 Aboiteaux *(dikes) in Grand-Pré, Nova Scotia, Canada,*
photographed in 1907. Insets show the ways in which aboiteaux *operated*
at low and high tides.

coast of India. Sandwiched between two rivers, the larger Ulhas that drains the Western Ghats and the shorter Thane Creek, most of the modern city lies on Salsette Island, but its older parts, further south, rise from what was once a series of seven islands. Known to the Greek geographer Claudius Ptolemy as the Heptanesia, knowledge of these islands underscores the freedom that today's Mumbaikars (and Indians more generally) enjoy, freed from past colonial yokes. But geography served a different purpose after these islands were acquired by King Charles II of England and leased in 1668 to the English East India Company, which took steps to join them together. Inspired by the land reclamations then happening in Holland (now the Netherlands) and the Fenlands of eastern England, the Company commenced a series of land reclamation works that eventually saw most of the islands joined to each other and to much larger Salsette Island. Subsequently during the British Raj, this became a common metaphor for unity not of islands but of the entire country of India under the British crown.[5]

Much like those in Acadia, the earliest reclamations at Bombay commenced with the building of rock-faced embankments (dikes) between islands with sluices allowing water to drain to the sea at low tide, permitting rice cultivation on the newly emergent lands. By the year 1715 when Charles Boone became Governor of Bombay, only one dike remained unbuilt, that needed to fill The Great Breach between Malabar and Worli islands. Expected to be completed in nine months, it actually took eight years. For it seemed that no matter how many boatloads of rocks were dumped into Worli Creek, the nascent embankment more often collapsed than rose. Until the day that Ramji Shivji Prabhu, an engineer on the project, dreamt that statues of the goddess Mahalakshmi and others lay buried in the creek mud and that, until these were disinterred and respectfully housed, the Breach would remain. The statues were located, a shrine built to accommodate them and finally the embankment was finished … and the five largest islands of the Heptanesia became one.

Like other coastal cities – you might think of Hong Kong, Macau and Singapore – subject to vast amounts of strategic land reclamation during their colonial histories, the reclamation of Bombay-Mumbai has left an unfortunate legacy that its sponsors could never have foreseen. Today, in an era of rising sea level, the reclaimed areas flood increasingly often and may be the first parts of this megacity to be permanently submerged, forcing their inhabitants elsewhere. The contemporary foundations of Mumbai's economic primacy in South Asia, which land reclamation helped establish, may be shaken as the frailty of these reclaimed lands is exposed by climate-driven changes over the next few decades.[6]

Today some of the largest, most successful and sophisticated land reclamation projects are found in the Netherlands. To understand why, we need to step back almost 2,000 years ago to the time when the people of this part of north-west Europe, notoriously low and flood-prone yet very fertile, began building dikes to keep the sea off the land. Dike construction was followed by the excavation of canals and, centuries later, the pumping of water from the land using the iconic windmills to create polders. Literally used to mean a piece of land higher than its surroundings, as in the modern Dutch word *graspol* (a tuft of grass), the word polder has come to be used more broadly today to refer to reclaimed land surrounded by dikes. Millions of Dutch people live on polders protected by thousands of kilometres of solidly engineered dikes, a situation understandably of growing interest to other nations where the livelihoods of coastal-dwelling peoples are under threat from rising sea level.

In the Netherlands, were these dikes not there, the sea would cover the polders to depths of many metres. Yet it could be said that the threat of Dutch dike failure occupies the minds of those who do not actually live in the Netherlands more than those who do (and whose faith in dike efficacy is ingrained). Of course, it is a small evidence base, but the famous story of Hans Brinker, a 'little Dutch boy' who saved the city of Haarlem by putting his finger in a hole in a leaking dike, was created in 1865 by an American children's writer Mary Mapes Dodge. The story is hardly known in the Netherlands. And if large dikes show signs of leaking and imminent collapse, the holes cannot be plugged with a finger – something bigger is required.

On the night of 1 February 1953, the unprecedented rise of water along the outside of the dike bordering the River Ijssel, which protected the most densely populated part of the Netherlands including the cities of The Hague and Rotterdam, was causing alarm. At 5.30 a.m. a 15m section of the dike gave way – the largest available ship, the *Twee Gebroeders*, was ordered to sail into the gap to plug it. It succeeded and its courageous captain, Arie Evergroen, had a statue erected in his honour in Nieuwerkerk aan den IJssel.[7]

Despite its obvious merits, dike building might not be a sustainable long-term strategy for adapting to rising sea level in the future, especially beyond the year 2100 when sea level is expected to continue rising whatever actions we take now to reduce greenhouse gas emissions. Yet at the same time it seems unlikely that people will no longer want to occupy coastal lowlands; our love affair with such places is pragmatic as well as immanent. There are lessons we can learn from history and the Netherlands is a good place to start.

As early as 650 BC, farmers settled the coastal regions of the modern Netherlands, specifically that of the Wadden

Sea (Waddenzee), drawn by the possibility of growing cereals as well as flax and water pepper in its salt marshes. But it was not simply a question of learning how to grow food in these regularly inundated wetlands, but also how to live there. The ingenious method developed by their first inhabitants was the construction of artificial earthen mounds named terps on which houses were built. Some terps were so large there was space for cattle byres as well as dwellings.[8]

A similar juxtaposition of adaptation (food and dwellings) comes from the Guianas (today Guyana, Suriname and French Guiana) on the north-east coast of South America sandwiched between the mouths of two giant rivers, the Amazon and the Orinoco. When the Dutch began colonising the deltaic Guianas early in the sixteenth century, they quickly saw how their pioneering skills in reclaiming swampy coastal lowlands might help transform what they considered the unliveable parts of these continental fringes into places that might become productive, justifying the expense of colonisation.[9] With this in mind, they began a programme of polderisation for the cultivation of sugar and later, coffee and cotton. Today few Guianan polders remain functional, most now overgrown with forest. But had the Dutch colonisers only looked more carefully, they might have noticed (as I am sure some did) evidence that the indigenous people of the Guianas had adapted to life in swampy coastal lowlands in much the same way as did the first settlers of the Atlantic fringes of the Netherlands more than 2,000 years earlier.

Research shows that, prior to European settlement of the Guianas, the inhabitants of the region had 'carefully organised, managed and anthropised their territory', largely through a system of raised fields and ditches, but also through the construction of mounds, the same as Dutch terps, on which houses could be built above the

level of the regularly flooded land all around.[10] In western
Suriname and eastern Guyana, the largest mound (named
Hertenrits or 'sandy ridge of the deer' in Dutch) comprises
a raised area of 40,000m², its construction requiring the
mobilisation of the equivalent of 14,000 truckloads of earth
moved by hand by the Arauquinoid people about AD 650.

The terp-centred society of the Wadden Sea coast and
the mound-building peoples of the Guianan wetlands
have something to teach us today. For so often do we hear
the mantra that the current era of climate change is
unprecedented and requires novel solutions, we might think
that history has nothing to teach us – but that is wrong.
Our ancestors learned to live in challenging environments
and to cope with environmental adversity in many forms
and in many places. The building of terps and mounds was
an adaptation strategy developed millennia ago that allowed
people to occupy and produce food in places once regarded
as uninhabitable. Small wonder then that terp construction
has been suggested as a way of continuing to occupy marshy
coasts in the future.[11]

As ever more mud was added to them, some Dutch terps
and Guianan mounds became so large that they merged
into one another, forming what might be better termed
islands. In this, they join more widespread traditions of
artificial island-building that characterised many cultures
at particular moments in their history.

Artificial islands are found all across the world, especially
in lakes and lagoons where they were never exposed to
destructive waves. For example, in lakes (lochs) across
Scotland, including the islands of the Outer Hebrides, are
found artificial islands named 'crannogs'. At first glance,
you would think them natural features, but research has

shown them to be clearly human-made, cleverly assembled from rock and wood. Some Scottish crannogs are 5,000 years old, the reasons for their creation still uncertain.

From what we know of their history – and the topic lacks systematic treatment – there seem to have been many reasons for artificial islands to be constructed, perhaps the commonest being as places of refuge. During times of conflict, being on land contiguous with that occupied by your aggressors invited attack, whereas living in a place surrounded by water, as did Hereward the Wake, created a barrier that might deter them.

It was flooding of valued coastal lowlands that prompted construction of the world's largest artificial island – Flevoland in the Netherlands, formed from the joining of the two Flevopolders. A series of storms in the first half of the twentieth century showed just how vulnerable much of the Netherlands was to flooding so a massive 30.5km-long dike – the Afsluitdijk (Barrier Dike) – was constructed to transform what had been a coastal inlet (Zuiderzee) into a freshwater lake named IJsselmeer. Once the IJsselmeer was isolated from the ocean, a series of dikes was built within it to create Flevoland, home today to some 370,000 people (Figure 14.2).

Another example of island creation comes from the Sacramento–San Joaquin Delta on the coast of California where European settlement began in the mid nineteenth century, simultaneously with the California gold rush. To both satisfy the escalating demand for familiar foods in San Francisco and take full advantage of the fertile delta soils, a patchwork of artificial islands emerged on which grains, potatoes and hay were planted, orchards grew, and cattle, sheep and horses grazed. Between 1850 and 1930, seventy-four islands were created involving 1,785km^2 of reclaimed land. The process was neither simple nor quick. Many dikes failed, nascent islands washed away. The building of

Figure 14.2 *Building the artificial island of Southern Flevoland near Muiderberg, 15 October 1963.*

Sherman Island took eight months to complete and involved the building of 65km, about 40 miles, of dikes made from stacking three-high cut-peat blocks that solidified when they dried out. Those early decades were times of hope, as this upbeat 1869 editorial from a local newspaper shows.

> In these reclaimable lands we shall have drought-proof means of life and luxurious living for the whole population of our State, were it twice as numerous. Heretofore the certainty of occasional famine years has been a dark cloud on the horizon before the thoughtful vision. Now we see salvation. All hail! to the great minds that have conceived this enterprise [of reclamation]. God speed their success and bring them rich reward.[12]

Things did not work out exactly as anticipated because keeping the sea off the land did not prevent it from being flooded by the river after heavy rain. Nor could seawalls and dikes prevent salty seawater penetrating the groundwater, especially in times of drought. About 60 years after it began, the accumulated problems associated with island creation in the Sacramento–San Joaquin Delta forced the state to take control of matters. Straightening of river channels to avoid flood impacts on peat dikes ameliorated the situation somewhat. Today the existence of dams in the upper reaches of the rivers mostly keeps flows manageable for the artificial-island dwellers. Yet the twin problems of subsidence and sea level rise are expected to pose future challenges to these islands, the surface of some of which today lie 3m below the average sea level.

On the other side of the Pacific Ocean, we find many examples of artificial-island building off the coasts of oceanic islands. Most examples are surprisingly under-researched and, to my mind, undervalued within the bigger picture of historical human endeavour. But think about it. An island, relatively small and remote, in the middle of an ocean. As its population grows, there comes a point when there is no further room for expansion. As the sea level rises or fragile coasts are periodically smashed by extreme waves, precious land is lost. This makes such islands ideal spots to consider creating more land.

The ninety-three artificial islands that comprise the megalithic site of Nan Madol off Pohnpei Island in the Federated States of Micronesia have long been a source of wonderment and speculation.[13] Just over 2,000 years ago, the earliest occupations at Nan Madol were on natural sand islets offshore. It is plausible to suppose that, as it became obvious these were too small to accommodate a growing demand, the artificial islands that exist here today were made. Built from boulders and sand, the larger islands at Nan Madol were topped with elaborate dwellings made from 'logs' of basalt rock, some weighing almost ten tonnes – we are unsure from where they came or indeed how they were mined and moved.[14]

Artificial islands have been constructed for other reasons. Inhabitants of human-made rock islands in the Langa Langa Lagoon off Malaita Island in the malaria-prone Solomon Islands (western Pacific) point to the absence of any mosquitoes in their breezy offshore homes as one reason for their construction. Some Malaitan traditions explain these islands were built for people to live on so they would not take up valuable agricultural land on the larger islands. The converse applies to the *chinampa* (floating food gardens) of Mexico and elsewhere in Latin America that were created in shallow lakes, using underwater fences to trap sediment, to allow food crops to be grown out of reach of shore-breeding pests.

In more recent times, construction of artificial islands has become a common solution to the twin problems of land loss and overcrowding, particularly in large cities: from the island of Dejima, built during the fifteenth century in the bay of Nagasaki in Japan to house Portuguese traders and keep them from contact with ordinary Japanese; to the artificial islands built for airports where there is simply no area of vacant land large enough to accommodate these. The artificial islands constructed off the coast of Dubai add significantly to the country's coastline; three islands will be shaped like date palms (Palm Jumeirah was completed in 2006), the fourth like all the world's continents joined together.

Today when we think what to do about rising sea level, we generally favour practical solutions like building sea defences or shifting valuable buildings and infrastructure out of danger zones – but some of us also pray. A spiritual response intended to engage a divine decision-maker and persuade them to halt the disruption. The tradition is thousands of years old, probably more than we shall ever know. And in the past, it is likely that spiritual responses were viewed as being at least as important as practical tangible ones in responding to a whole range of external, seemingly unmanageable forces like sea level rise – and drought and flood and disease.

While we know (or at least infer) something about our distant ancestors' practical responses to the rise in post-glacial sea level, we know almost nothing for certain about their spiritual responses – other than that they undoubtedly existed and were invested with great importance. But we can speculate. For example, from more than 7,000 years ago in Australia, Aboriginal stories have reached us which recall times when people, alarmed by how fast the ocean was drowning the fringes of the land and fearful it might be completely submerged, built wooden barriers to check its progress.[15] The stories are silent about any spiritual response these people might have made but, along the modern coasts of Australia, which are about 7,000 years old, we sometimes find stone arrangements that may have been intended for the 'ritual control' of rising sea level and extreme waves.[16] This may seem to you like a stretch. After all, what is a 'stone arrangement'? Well, it is something out of place, clearly planned not arbitrary; something demonstrably ancient; something local traditions may suggest to be invested with meaning. But perhaps most persuasively, such stone arrangements are found not just in Australia but along much of the world's coast so we might deduce they are a common expression of ancient societies. But of what?

Along the Atlantic coasts of north-west Europe where unusually, as explained in Chapter 8, sea level has been rising continuously since the last ice age ended, the occupants of now-submerged cities like Cantre'r Gwaelod and Ys sought to extend their lifespan by the construction of sea defences. Overwhelmed eventually, these physical remedies would have been accompanied by spiritual ones. What these were is almost impossible to reimagine, you might think, although a bold attempt by a group of French scientists makes compelling reading. They studied the stone lines at Carnac, for me one of the wonders of our world. Understated, often overrun with grazing sheep, thousands of standing stones (or menhirs) extend in parallel rows for many kilometres across the landscape and even beneath the ocean surface. When first they came to the notice of western science, the stone lines were thought to be astronomical alignments (they are not), perhaps markers for graves of fallen warriors (no), maybe even land boundaries (also no). The novel idea of scientists like Agnès Baltzer and Serge Cassen is that these standing stones were arranged to create a barrier between the material world and the metaphysical one, a cognitive barrier intended to prevent interference by the gods with the human world, specifically to stop the rise of the ocean across the land. All this is thought to have occurred 6,500 years ago, to represent deductive reasoning by people living in the Carnac area at a time when there was a short-lived acceleration in the rate of sea level rise here that prompted desperate measures.[17]

It seems the response failed. Sea level continued rising here, its most emphatic retort being the submergence of parts of these stone lines. From where you sit now, you might think it was blindingly obvious that building stone lines to support an invisible mesh to stop deities interfering with earthly affairs would fail. But today we have little inkling about what people's belief systems entailed 6,000 years ago in this part of Europe. What we can say, given the phenomenal effort involved in creating these stone lines, is that there must have been widespread and enduring belief in their likely efficacy.

Such efforts were not isolated although we are only just starting to glimpse how our ancestors, thousands of years ago, rationalised and responded to the rise of sea level. One clue lies in where and how they buried their dead. Many ancient cemeteries were located on coastal fringes, often on small offshore islands, some even in places that were submerged at high tide. Why would anyone choose to bury dead bodies along the very edges of the land? The answer could be pragmatic – it was unvalued land, fit for little else – but more recent thinking on the matter has inclined more towards the idea that our distant ancestors viewed the coast as a place of transition between earthly and spiritual worlds. This is precisely what underpins Pacific Islanders' traditions of dead souls jumping off the ends of the land (see Figure 5.1) and may explain a whole range of beliefs about the ocean, touched upon in Chapter 4, that have been almost completely forgotten today. If you believe that the ocean covers the place where you will spend your life after death, what better spot to lay a dead body than right next to the ocean, along with all the items the dead person needs in the afterlife?

Yet, however compelling this explanation might sound, its simplicity troubles me. We may be missing the bigger picture. One hint at what this might involve is the novel explanation for the stone lines at Carnac as being (or supporting) a cognitive barrier designed to halt rising sea level more than 6,000 years ago. Another hint lies in the extant stories of Aboriginal Australians recalling the anxiety they felt at the inexorable rise of post-glacial sea level and the possibility it might engulf the entire land; a thought that prompted decisive action. These two examples suggest to me that coastal peoples the world over felt threatened, far more than we realise, by the eight millennia or so of rising sea level they experienced in the aftermath of the last ice age, in most places 15,000–7000 years ago. Might that threat have weighed so heavily on coastal dwellers that they developed stories about the ocean-as-aggressor, as a swallower-of-land, and deified or demonised it? And having given it divine status, might people have then sought ways to appease it, to try and stop it drowning the land?

It has been proposed that certain objects commonly used by coastal peoples thousands of years ago were invested with symbolic meaning and power. An example are the fish hooks made from elk bone used 10,000 years ago by the Mesolithic inhabitants of the Skagerrak coast of Norway and Sweden. The elk was a much-admired animal, not least because it could live on the land yet was also able to swim great distances across the ocean, dive and feed on water plants. It was considered a liminal creature able to cross the boundary between land and sea more effortlessly than humans, which may have suggested to them it had a privileged relationship with the ocean, even able to intercede on behalf of humans and quell the ocean's periodic bursts of fury that made fishing and travel impossible. This may be why the Skagerrak people once favoured elk bone for carving fish hooks, not solely as functional tools but also as ones imbued with power, able to help their human users negotiate with the sea.[18]

From here it is only a short step to explain why ritual burials of such material are found along many coasts. In the Netherlands at the Hoge Vaart site, accidentally discovered during motorway construction, there are burials of flint tools alongside the skulls of aurochs within a former coastal wetland that may indicate such an association during Neolithic times, slightly more than 6,000 years ago. And then in various parts of the British Isles are found even older coastal middens containing a range of similarly suggestive material perhaps, in the words of one scholar speaking of those along the fringe of Oronsay Island, referencing a 'textural boundary ... between land and sea'.[19]

In eastern England in the year AD 624, almost 1,400 years ago, the *Bretwalda* (or high king) named Raedwald died. So powerful had he been within the Anglo-Saxon lands of central and southern England that many lesser monarchs came to his funeral bearing lavish gifts. As was common at the time when important people died, a barrow (or earthen mound) was erected over Raedwald's body, but it was what was found in 1939 inside this barrow, untouched by grave robbers, that makes this story so extraordinary. Raedwald's mound is at a place called Sutton Hoo east of Ipswich and a couple of kilometres from the coast in AD 624. It seems his mourners dragged a large ship onshore in which to bury Raedwald along with those of his possessions he valued most and many of the gifts presented at his funeral. When this was over, everything was encased in earth, creating a giant mound (Figure 14.3).

Regarded as unique at the time it was discovered, the Sutton Hoo ship burial is actually far from this. Ship burials

Figure 14.3 *The ship burial at Sutton Hoo, eastern England, showing (A) details of the 1939 excavation, (B) the type of Anglo-Saxon long ship interred here and (C) the helmet of King Raedwald.*

of important people have been found in many parts of north-west Europe involving both Anglo-Saxon leaders like Raedwald and also Vikings. On the island of Saaremaa in Estonia, two Viking ships have been excavated and found to contain the remains of at least 42 warriors, possibly aggressors defeated by the Oeselian islanders and magnanimously given a ritual ship burial.

It may be that ship burials of the kind Raedwald was given were not simply appropriately grand and stately, but also were considered to ensure that the dead person would be able to return promptly to the ocean's embrace, equipped with everything a person of high status would need there. As Tennyson wrote of the death of King Arthur after he was pushed offshore in the barge that would take his broken body to the island of Avilion, a mystical place 'crown'd with summer sea', the onlookers on the shore watched 'till the hull look'd one black dot against the verge of dawn'.[20]

Today we regard the coast differently because we regard the ocean differently to the ways our ancestors did long ago. The ocean is no longer the mysterious depthless place of beginnings and endings. It is not a place we need to regard as a source of aggression for with our understanding – our ability to scan the ocean from top to bottom, side to side and to monitor its every ripple – we have nothing to be apprehensive about. But that is something that has come at a very late stage in human history.

So next time you are wandering a deserted stretch of coastline and you come across a scatter of boulders, imagine what they might represent. Perhaps the remains of an attempt to stop the apparently unremitting encroachment of the sea on to the land. Perhaps a reminder that off this particular coast there once stood places – villages, towns and cities – that people once inhabited, places now forgotten or at best stirred into the soup of folklore and mythology. Perhaps a reminder that there is so much more to be learnt about human history and that much of what we do *not* know lies below the ocean surface. Yet pinch yourself and know that you are real. For the very fact that you and I have made it this far, across 200,000 years of time punctuated by occasional extreme, almost unimaginably large and disruptive events, shows that humanity is resilient and able as a species to survive just about anything nature cares to throw at us.

Further Reading

Benjamin, J. and others, editors. 2011. *Submerged Prehistory*. Oxbow, Oxford.

Cunliffe, B. 2001. *Facing the Ocean*. Oxford University Press.

Darwin, C. R. 1870. *A Naturalist's Voyage Around the World*. John Murray, London.

Evans, J. G. 2003. *Environmental Archaeology and Social Order*. Routledge, London.

Gaffney, V. and others. 2009. *Europe's Lost World*. Council for British Archaeology, York.

Gillis, J. R. 2004. *Islands of the Mind*. Palgrave Macmillan, New York.

Gillis, J. R. 2013. *The Human Shore*. Chicago University Press.

Higginson, T. W. 1898. *Tales of the Enchanted Islands of the Atlantic*. Macmillan, New York.

Hsu, K. J. 1983. *The Mediterranean was a Desert*. Princeton University Press.

Kelly, L. 2015. *Knowledge and Power in Prehistoric Societies*. Cambridge University Press.

Kelly, L. 2016. *The Memory Code*. Allen and Unwin, Sydney.

Kerr, R. 1824. *A General History and Collection of Voyages and Travels*. Blackwood, Edinburgh.

Menzies, G. 2004. *1421: The Year China Discovered the World*. Bantam, London.

North, F. J. 1957. *Sunken Cities*. University of Wales Press, Cardiff.

Nunn, P. D. 1994. *Oceanic Islands*. Blackwell, Oxford.

Nunn, P. D. 1999. *Environmental Change in the Pacific Basin*. Wiley, Chichester.

Nunn, P. D. 2009. *Vanished Islands and Hidden Continents of the Pacific*. University of Hawaii Press, Honolulu.

Nunn, P. D. 2018. *The Edge of Memory: Ancient Stories, Oral Tradition and the Post-Glacial World*. Bloomsbury, London.

Oppenheimer, S. 1998. *Eden in the East: The Drowned Continent of Southeast Asia*. Weidenfeld and Nicolson, London.

Ramage, E. S. 1978. *Atlantis: Fact or Fiction*. Indiana University Press.

Ramaswamy, S. 2004. *The Lost Land of Lemuria: Fabulous Geographies, Catastrophic Histories*. University of California Press, Berkeley.

Stommel, H. 1984. *Lost Islands*. University of British Columbia Press, Vancouver.

Notes

Details of key books referred to in these notes may be listed separately under *Further Reading*, others referenced directly. Where citation of other sources, usually scholarly articles, is essential then only their first author, their year of publication and the journal name (in italics) are given.

Chapter 1: Introduction: Hearing the Past

1 The two quotes are from Barbeau's 1995 biography.
2 This story may recall the submergence of Haida Gwaii during the Kinggi period 9,700–12,700 years ago. I have paraphrased this particular Haida story using Barbeau's own words with additional details from his biography. In the original story, Henry Young also told Barbeau that 'many of our people know this story … for it is the truth. It really happened, a great many years ago'. A comparable situation from Triquet Island further south in the same area involves Heiltsuk traditions recalling that Triquet was never ice-covered (like much of the surrounding area) and was thus a place to which the Heiltsuk returned to repeatedly within the last 14,000 years. Speaking for the Heiltsuk Nation in 2017, William Housty noted how 'to know that that history has survived thousands of years of [oral] transmission through different generations to the present day, and to have that reaffirmed by archaeological information, is very powerful … the history that we've been talking about for thousands of years is true, it's not just something that has been made up' (*Times Colonist*, 16 April 2017).

Chapter 2: Worlds in Shadow

1 This is not to say there have been no efforts to raise popular awareness about submerged lands. Among my heroes is the trailblazing Nic Flemming, whose seminal work is the 1972 book *Cities in the Sea*, focused on his exciting underwater discoveries in the Mediterranean.
2 During the coldest time of the last great ice age, because sea level was much lower than today, the total area of land in North America was 31 per cent greater than it is today. According to the 2018

calculations of Joanna Gautney (*Journal of Archaeological Science, Reports* 19), ice-age North America had a total land area of 35.8 million km2 while today it occupies 24.7 million km2, a difference of more than 11 million km2. For the purposes of constructing Figure 2.1, it is calculated that the area of the conterminous United States is 8 million km2 and that the combined area of the (missing) States of Arkansas, Iowa, Kansas, Louisiana, Minnesota, Missouri, Nebraska, North Dakota, Oklahoma, South Dakota and Texas totals 2.5 million km2 or approximately 31 per cent.

3   The details of the story about Dahut and her eventual demise are described in my 2018 book, *The Edge of Memory*, in which I propose that the submergence of Ys could have occurred 8,750 years or more ago. Stories about Ys were collected and analysed at length in the 1926 book by Charles Guyot, translated into English as *The Legend of the City of Ys* and republished in 1979.

4   All the quotations and most details in this section come from the two chapters devoted to Morrell in the superbly instructive 1984 book by Henry Stommel.

5   My 2003 study in the *Annals of the Association of American Geographers* argues that many aspects of the Maui myths are attributable to observations of shallow-water volcanism.

6   My 2004 study in the *Journal of Pacific History* reviews the geological meanings of Niuean myths and comes to much the same conclusion.

7   In the *Critias*, Plato wrote that Atlantis was an island 2,000 stadia by 3,000 stadia. One stadia is roughly 200m, so Plato's Atlantis was supposedly about 240,000 km2, about the size of the UK or the US state of Michigan.

8   For example, in his 1978 book *Atlantis: Fact or Fiction*, E. S. Ramage concludes that 'what Plato seems to be doing with the *Timaeus* and *Critias* … is putting some of his theorising of *The Republic* on a more practical level by giving an example of how it might work in practice' (p 20).

9   It is at this point, of course, that the pseudoscientists start jumping up and down, protesting that Atlantis has been found – many times, in fact. And therein lies the rub. So many times in different places has it been claimed Atlantis has been found that most of those claims must be false. The test for geologists is scientific proof. And there is none of that when it comes to Atlantis although, as shown in this book, there are memories and myths of submerged lands in many parts of the world – just none meeting the description of Atlantis. Not that this would be expected anyway.

10   More detail about the sources Plato used to ground his story of
     Atlantis in fact (and beguile later readers) are described in my
     2009 and 2018 books.

Chapter 3: Recently Drowned Lands

1   From p. 414 of Arber's 1910 edition of the *Travels and Works of
    Captain John Smith*. After the storm subsided, 'repairing our fore
    saile with our shirts, we set saile for the maine'.
2   Paraphrased from the account in S. W. Hall's 1939 book entitled
    Tangier Island. For a contemporary picture of life on dwindling
    Tangier Island, I recommend the lively 2018 account, *Chesapeake
    Requiem*, by Earl Swift.
3   Quoted in *The Virginian Pilot* on 3 August 2017.
4   From an article in *Chesapeake Quarterly*, October 2014, by Rona
    Kobell.
5   The words of Lorie C. Quinn Jr quoted in Mindie Burgoyne's
    2009 *Haunted Eastern Shore*.
6   The reason why Chesapeake Bay is sinking is explained in
    Chapter 8. In short, because it was the site of a crustal bulge created
    during the last ice age by the weight of the ice sheet to the north,
    it is now sinking as this bulge is collapsing (as shown in Figure 8.2).
7   One version of the Sipin story is in the 1983 compilation of Micronesian
    stories, *Never and Always*, by Gene Ashby. Another was told to me and
    my family by the late Yapese historian, John Runman, one sultry
    October afternoon in 2014 on a beach near the village of Malaay.
8   With some colleagues, I reported and reviewed some such stories
    from Vanuatu in a 2006 article in the *Journal of the Royal Society of
    New Zealand*.
9   I collected stories about Solo first hand when I travelled there with
    people from nearby Dravuni Island. All long before I read Thomas
    Westropp's 1912 account of the submerged city of Kilstapheen off
    the coast of County Clare in Ireland 'overwhelmed by an irruption
    of the sea long ago' (*Proceedings of the Royal Irish Academy*, p. 250).
    Westropp collected stories from local people who stated that the
    remains of the city could sometimes be seen by those sailing over
    them and that the enchanted inhabitants of Kilstapheen might
    raise 'destructive storms over it while the waters were calm
    elsewhere', something similar to that threatened at Solo.
10 The story from the Isle of Man comes from p. 481 of F. S. Bassett's
    1892 book, *Sea Phantoms*. That about the inn at Singleton Thorpe
    comes from the 1837 book by W. Thornber on *The History of Blackpool*.

11 Most details about Burstall Priory come from the 1912 book, *The Lost Towns of the Yorkshire Coast*, by Thomas Sheppard. A stone doorway from Burstall Priory is now in Easington Church. The sketch in Figure 3.2 is on p. 497 of the 1841 book about Holderness by George Poulson.

12 This information comes from the commentaries for a 2002–03 exhibition at the National Museum of Denmark (Nationalmuseet) entitled *Mare Balticum*.

13 Maybe Vineta never was. In the 2003 collection (*Baltic Sea Identities*) by J. Litwin, Michael Andersen argues that the story of Vineta 'reflects the need for a myth and a mythical past' (p. 24) analogous to the role of Atlantis in the cultural history of the Mediterranean.

14 Quotes from the study by John Broich (2001, *Environment and History*).

15 Prior to the completion of the Aswan High Dam in 1970, an average of 85.1km3 of freshwater and 120 x 106 ton of suspended solids poured into the Nile Delta every year (Williams, 2000, *Global and Planetary Change*).

16 According to the account in Homer's *Iliad*, the Greek general Odysseus came up with the idea that led to the siege's end. He created a giant hollow wooden horse, an animal sacred to Trojans, which he filled with warriors. Then razing his camp and setting sail with his army, Odysseus led the Trojans to believe the siege was over. Considering it a trophy, they moved the horse into the city and, after its inhabitants fell asleep after a night of celebration, the Greek warriors came out of the horse, opened the gates of the city and started the massacre of its occupants.

17 This does seem true to form for Paris, who in Homer's *Iliad* is represented as cowardly by nature and less than honourable for having killed the Greek hero Achilles with a blow from behind.

18 The quote comes from Macaulay's 1914 translation of *Herodotus*.

19 A similar comment can be made about the city of Troy which, until the excavations by Heinrich Schliemann in 1870, was considered by many never to have existed.

20 The quote comes from Arnold's 1858 translation of the *Euterpe of Herodotus*.

21 Both this and the previous quote from Strabo's *Geographica* come from the translation in Volume 8 of the 1932 Loeb Classical Library edition.

22 The credit for this belongs to the team headed by Franck Goddio who has written several books on the subject including the excellent *Sunken Cities: Egypt's Lost Worlds* with Aurélia Masson-Berghoff in 2016.

23   The weight of sediments in large deltas like the Nile is so great that
     it can change the shape of the Earth's crust beneath, similar to what
     happens when you half-fill a balloon with water. Subsidence of
     deltas is also due to the progressive compaction of delta sediments.
     Various parts of the Nile Delta have been subsiding at rates of
     between 1 and 5 mm per year for several thousand years. Data from
     Port Said indicate an average subsidence rate of 4 mm per year over
     the past 8,540 years. When pondering future submergence of the
     Nile Delta, these rates must be added to projections of future sea
     level rise, the result being that within the next century there will be
     profound and largely unavoidable changes that will require not
     only out-migration of people from its lower parts but also a
     significant realignment of livelihood activities across a vast region.
24   Most of the information about the archaeology of Pavlopetri here
     comes from Chapter 17 by Jon Henderson and others in the 2011
     book edited by Jonathan Benjamin.
25   In 464 BC, an earthquake produced by thrusting along the Sparta
     Fault flattened the Peloponnesian city of Sparta and had massive
     societal ramifications. Research shows this was not an isolated
     incident and that this fault ruptures every 500–1,000 years
     (Papanastassiou, 2005, *Journal of Geodynamics*). It is possible that
     either the 464 BC event or those in AD 550 or AD 1000
     contributed to the abrupt (coseismic) subsidence of Pavlopetri.
26   The key research was reported in the 2015 issue of the journal
     Tuberculosis by I. Hershkovitz and others. In a later paper by Helen
     Donoghue and others (2017, *Diversity*), aDNA and lipid biomarkers
     were used to confirm human tuberculosis in the 9,000-year-old
     bones of a woman and child living at Atlit-Yam (and in a
     2,700-year-old Egyptian mummy). Understanding the ancient
     roots of tuberculosis can help reduce its spread.
27   Many of the human remains recovered from Atlit-Yam showed
     an ear condition (alternobaric vertigo) brought about by routine
     immersion in cold water.
28   The dietary diversity of people living at Atlit-Yam nine millennia
     ago has been determined directly by isotopic analyses of bones
     and food residues, but also by identifying plausible uses for stone
     tools found there.
29   The link between a flank collapse of Mount Etna and the end of
     Atlit-Yam was cogently argued by M.T. Pareschi and others
     (2007, *Geophysical Research Letters*). It demonstrates how
     catastrophic collapses of active volcanoes can affect places far
     away, something these authors argue should be acknowledged in

modern risk planning for Mediterranean coastal cities. Similar arguments apply to future collapses of Atlantic island volcanoes like Cumbre Vieja in the Canary Islands (see Chapter 13) and to oversteepened delta fronts like that of the Magdalena River in Colombia (see Chapter 13).

30 Charles Darwin writing to Arthur Russel Wallace on 22 December 1857.

Chapter 4: At the Nexus of Science and Memory

1 'That Country by a deluge swallow'd – That martyr'd land, Cantrev y Gwaelod, So sudden from existence sunder'd' (extract from Canto I of the 1824 poem 'The Land Beneath the Sea' by T. Jeffery Llewelyn Prichard).

2 For most of the time it existed, this story would have been known only orally. An account from 1804 is one of the earliest to be written down. 'About the year 500, when Gwyddno Goronhir was lord of this hundred, one of the men who had care of the dams got drunk, and left open a floodgate. The sea broke through with such force as also to tear down part of the wall, and overflow the whole hundred, which, since that time has always been completely flooded,' (Bingley, 1804, *North Wales*).

3 For example, the work of Erin Kavanagh and Martin Bates (2019, *Internet Archaeology*) avers that the perspective of the storyteller might somehow negate the links that can be drawn between such stories and past sea level changes. This is certainly true of Plato's Atlantis for we know what Plato was trying to do with his fictional narrative, but I cannot imagine why this would be true of every other story about submerged lands.

4 Archaeological evidence suggests people were living in what we now call Ireland at least 12,500 years ago before rising sea level drowned the land bridge connecting it to the rest of Europe at the time. Are memories of this time preserved in ancient stories? One possibility is the story of Fintan, a merman living on the coast of Ireland before it became an island and who is said to have become human after the subsequent 'deluge', allowing him to live on the land (and eventually become a saint).

5 In his widely quoted 1957 book *Sunken Cities*, F. J. North argued that the story of the submergence of Cantre'r Gwaelod, which he believed to have involved 'a sudden rising of the sea or a sinking of the land' (p 149), could not be true simply because 'such a phenomenon [is] unknown in geological history' (p 150).

This may have been viewed as a watertight argument (by historians) in 1957 but certainly not today when the precise nature of sea level changes during and after the last great ice age has been known for decades.

6   An account of the lands around Cantre'r Gwaelod, the Lowland Hundred, suggests they were 'replete with smiling farms and neat villages' (North, 1957, p 156).

7   Extract from the writings of the Tamil scholar A. M. Paramasivanandam; translations in the engrossing 2004 book by Sumathi Ramaswamy.

8   See pp. 16–17 in my 2009 book.

9   Details in this section come from the 2004 work of S. C. Jayakaran (*Indian Folklore Research Journal*).

10 This extract is from Paramasivanandam; translation in Ramaswamy (2004). Previous quote in this paragraph from p. 152 of North (1957).

11 The arguments are laid out in my 2018 book, *The Edge of Memory*, in which evidence is described for coastal-drowning stories being more than 10,000 years old, evidence for stories about volcanic eruptions being at least 7,600 years old and evidence for stories about meteorite falls being 4,200 years old – and many others. To understand how such memories were preserved in non-literate contexts, I recommend Lynne Kelly's 2016 book, *The Memory Code*.

12 This is from p. 15 of the 4th corrected edition of Martin's *A Voyage to St Kilda*, published in 1753. The key page is shown in the colour picture section. Intriguingly, 'there exists in the traditions of Harris [Island] a long and involved legend of a female warrior who used to hunt over the dry land between the Long Island [Harris-Lewis] and St Kilda' (pp. 27–8 in the 1972 book *Island on the Edge of the World* by Charles Maclean).

13 The ocean floor between St Kilda and Harris is mostly no more than 50–60m deep although there is a 4km-wide passage close to the former that is more than 100m deep. Owing to crustal rebound (illustrated in Figure 8.2) in this area, the reconstruction of past (post ice-age) sea levels is not straightforward. While some models suggest that sea level was 50–60m lower around 8,000 years ago, more recent work suggests sea level never fell this low after the ice in the area melted (Jordan, 2010, *Journal of Quaternary Science*).

14 In his 1980 opus, *The Languages of Australia*, Robert Dixon notes how 'many tribes along the south-eastern and eastern coasts have stories recounting how the shoreline was once some miles further out; that it was ... where the [Great] barrier reef now stands' (p. 46).

15 The taboo that Gunya broke was to catch a forbidden fish, a stingaree in some versions of the story. As punishment the ocean began to rise and cover the land. The water lifted up Gunya's canoe, carrying it rapidly inland, past Wunyami (Green Island) until it was beached near modern Cape Grafton. There, sheltering in a large cave, to their astonishment, Gunya and his companions found people from many other tribes … 'The cave is a camping place. They came and found people there, lots of people sitting there … Gunya spoke to the Gurragulu people, using names in their Gungganyji language. And Mayaar Yidinyji men … and Ngajanji people too. All these people stopping here in the one camp … they had all been driven to high ground by the rising sea waters' (from the story told by Dick Moses at Yarrabah on 22 August 1973, from pp. 92–93 Robert Dixon's 1991 book *Words of Our Country*).

16 From the pattern of cut branches on the lower parts of the trees, Tasman's crew thought the invisible inhabitants of Tasmania might be giants (rather than adept climbers), an idea that stayed current for a remarkably long time among early European visitors to Tasmania and later to New Zealand.

17 The dingo was a much later arrival than the first Australians. The latter arrived at least 65,000 years ago, while the dingo, likely to have been introduced by Toalean hunter-gatherers from Sulawesi (Fillios and Taçon, 2016, *Journal of Archaeological Science*), arrived a mere 3,300 years or so ago, as determined from the direct dating of dingo bones by Jane Balme and others from a cave in the Nullarbor Desert (2018, *Scientific Reports*).

18 The Gippsland 'drowning' story quoted here is estimated to be between 9,240 and 12,460 years old (Nunn and Reid, 2016, *Australian Geographer*).

19 The original story is in Thiele's (1843) *Danmarks Folkesagn*.

20 Quotes are from pp. 165 and 172 of *The Pirate*. A comparable real-life incident was reported from the River Danube (Austria) by Auguste Ellrich; it involved a man falling overboard from a boat in this wide river. After a lengthy struggle he drowned, his comrades on board having made no effort to save him. 'While he was battling among the impetuous waves, the crew stood quite composedly on deck, and cried out in chorus, 'Jack! Jack! Give in – dost thou not see that it pleases God?" (Jones, 1880, *Credulties*).

21 Most details in this section are from the study by Gerardo Pappone and colleagues (2019, *Geosciences*).

22 The details of Norris's discovery, outlined by Peter Baker (1967, *British Antarctic Survey Bulletin*), are compelling. Bouvet Island was

called 'Liverpool Island' by Norris who describes it in much the same way as it is today. His account of Thompson Island, 15 leagues north-northeast of Liverpool Island, is detailed and precise. In addition, Norris shows a sketch of Thompson Island that could not – from any angle – have been Liverpool/Bouvet.

23   Actually, this remains the most parsimonious interpretation of the available evidence. Representative of this view is Henry Stommel who in his trailblazing 1984 book *Lost Islands* labelled people who believed in Thompson Island as 'gullible' (p. 99).

24   This paragraph is based on research by pioneering historical climatologist H. H. Lamb, responsible for the first global survey of climate history from direct (human-recorded) rather than proxy (like tree-ring) records. His study of Thompson Island appeared in the 1967 *British Antarctic Survey Bulletin*.

25   The announcement in the House of Commons must have been during or before 1928 for the quotation is given in an anonymous article that year on p. 546 of *The Geographical Journal*.

Chapter 5: Red Herrings: Fishy Tales of Unlikely Sunken Lands

1   Nicholas Humphrey put it neatly in his 1996 book, *Leaps of Faith*, when he wrote that 'all great supernatural belief systems – indeed all great philosophical systems, up until now – have catered to two central needs: the need for a rational understanding of the surrounding world, and the need for emotional security within it' (p. 10).

2   Unfortunately, many abandon the task, 'baffled by a cloud of woolly counter-counter-claims hurled back by the believers' (from p. 48 of Nic Flemming's chapter in the 2006 collection of *Archaeological Fantasies* by Garrett Fagan). The earlier quote in this paragraph is from p. 42–3 of the opening chapter of the critical 1978 study of Atlantis by E. S. Ramage.

3   Beware indeed of professors, bespectacled or not! In his 1885 book *Paradise Found*, which ran through eight editions in a year, the Reverend Dr W. F. Warren, one-time President of Boston University, averred that the 'cradle of the human race ... was situated at the North Pole, in a country submerged at the time of the Deluge' (p. 47). Neither assertion is even remotely correct.

4   Plato's greatest disciple, Aristotle, reached the same conclusion, noting that Atlantis existed only in Plato's imagination; 'he who created it also destroyed it', he quipped.

5   From p. 276 (Chapter VII) of Volume 2 of the 1903 Yule-Cordier edition of *The Book of Ser Marco Polo*.

6   This is the 1889 book *Natürliche Schöpfungsgeschichte*, translated as *The History of Creation*. I consulted the 8th edition.

7   This judgement of Haeckel's motives may be unduly harsh. It may be simply that Haeckel strayed out of his area of expertise and came up with an idea he naively considered reasonable, never imagining that it would be so improperly used subsequently. Another scientist who may have had this experience is Masaaki Kimura, the geophysicist whose ideas about the human origins of undersea structures like the Yonaguni Monument in southern Japan have made him the darling of pseudoscientists.

8   The quote is from p. 172 of volume 2 of Blavatsky's 1888 opus, *The Secret Doctrine*. An absorbing and grounded account of the roots of Theosophy is Peter Washington's 1993 book *Madame Blavatsky's Baboon*.

9   Upon reaching the islands of Hawaii and finding their people spoke a language similar to those of Tahiti and New Zealand, Cook asked 'How shall we account for this Nation spreading itself so far over this vast ocean?'. Pacific scholar Damon Salesa argues that Cook could not have failed to have been hugely impressed by the abilities of the Pacific Island seafarers he encountered. He saw ocean-going vessels (longer than his own *Endeavour*) that could accommodate 150 people. He adjudged the Tahitian *pahi* (catamaran) as able to sail almost 200km a day. Yet Cook continued to disparage the maritime abilities of Pacific Islanders referring, for example, to their vessels merely as 'canoes'. Like most European explorers of the 18th and 19th centuries, Cook suffered from 'an inability to come to terms not so much with what Polynesians knew but … how they knew it' (Damon Salesa, used with permission).

10 The three places where ancestral Pacific Islanders are thought to have reached the western seaboard of the Americas are in California, Panama and Chile. The Chumash people who traditionally occupied the area around modern Santa Barbara in California are thought to have had contacts with Pacific Islanders, probably from Hawaii, which enabled the building of sewn-plank canoes (*tomolo*) (Jones and Klar, 2005, *American Antiquity*). Between the years 1516 and 1539 when the Spanish first explored Panama, they found numerous groves of coconut – a plant originating in the south-west Pacific – that could have reached there only if people had brought it by boat across the west-flowing Humboldt Current, perhaps from the islands of French Polynesia (Ward and Brookfield, 1992, *Journal*

of Biogeography). And finally, most compellingly, caches of 'Polynesian chicken' bones have been found at El Arenal in southern Chile, the remains of meals that could have only been provided by Pacific Islanders (Storey and others, 2007, *Proceedings of the National Academy of Sciences USA*).

11   Ghost islands abound in the Pacific. Several are mentioned in the sailing directions of traditional navigators. A warrior from the island of Pohnpei is said to have once visited the 'mythical' island Kachaw, bringing back a new variety of mangrove that became invaluable for housebuilding (Geraghty, 1993, *Journal of the Polynesian Society*). An analogy from north-west Europe is Mona, the 'moving island' off the coast of the Isle of Man.

12   Sometime between 1914 and 1926 when he was visiting the Kiribati island of Makin-Meang in his capacity as District Officer, Arthur Grimble heard many stories about ghosts. In the particular part of the island where dead spirits were said to tramp their way to paradise, Grimble accompanied by the local policeman saw a man limping towards him. Outraged when the limping man ignored him completely and walked by, Grimble made his way to the nearest village and described the man to the native magistrate. He knew at once who it was and took Grimble to see his dead body in a nearby house. The limping man had died shortly before Grimble saw him walking past.

13   One story from Kiribati, related in Chapter 9 of my 2009 book, explains what might have happened to American aviator, Amelia Earhart. It is said that as her plane was faltering, she saw from the air a submerged island named Bikenikarakara and landed there, believing it to be real.

14   This translation is from Andersen (1969), modified by Dr Pila Wilson.

15   The term 'medieval islomania' was introduced in the captivating 2004 book *Islands of the Mind* by John Gillis to describe the preoccupation with imagined islands in the Atlantic during the Middle Ages.

16   Purists should note that I spell the saint's name as Brandan because that is the way it was first written down, on the 1275 *Hereford Map* described by Benedict (1892, *Bulletin of the American Geographical Society*).

17   The quote is from pp. 34–5 of the 1922 book, *Legendary Islands*, by William Babcock. Later developments brought reports of an island (Isla San Borondón), now apparently vanished, in the far west of the Canary Islands that was equated with St Brandan's Island.

Although no one ever landed on the island, it was apparently seen on several occasions in this part of the Atlantic. On 3 May 1759, a Catholic priest named Viere y Clarijo saw it and sketched it in the presence of more than 40 witnesses. The form suggests that it was a representation of La Palma Island in the group which had somehow become reprocessed as an ocean mirage.

18   From p. xv of the 1898 book by Thomas Higginson.

19   The earliest cogent argument for European crossings of the Atlantic predating Columbus is that of Babcock (1917) in the *Proceedings of the International Congress of Americanists*. The intriguing idea that maps of the coast of the Americas made by Chinese voyagers who visited there in 1421 found their way into European map collections and facilitated the first trans-Atlantic voyage by Columbus was presented by Gavin Menzies in 2004.

20   Quoted from p. 139 of the account by O'Flaherty in the 1825 issue of the *Transactions of the Royal Irish Academy*.

21   The description is by Captain James Newton of the busse *Emmanuel* quoted on p. 183 of Payne's (1893) edition of *Voyages of the Elizabethan Seamen to America* (Hakluyt Society).

22   The quote is from p. 177 in Babcock (1922).

23   Quote is from Chapter 9 of the online version of the (1911) book *Folklore of Clare* by Westropp.

24   A powerful expression of this was given by W.H. Auden in his 1933 poem 'Paysage Moralisé' –

… *hunger was a more immediate sorrow,*
Although to moping villagers in valleys
Some waving pilgrims were describing islands …
'The gods', they promised, 'Visit us from islands,
Are stalking, head-up, lovely, through our cities;
Now is the time to leave your wretched valleys
And sail with them across the lime-green water …

25   See Handy's (1930) compilation of *Marquesan Legends* published by the Bishop Museum in Hawaii. To understand the fatalism of the pre-modern inhabitants of remote islands, it helps to read the writing of Sir Raymond Firth about the comparatively high incidence of suicide in the early 20th century on Tikopia, an uncommonly remote island in the eastern Solomon Islands. Many suicides were not explicit, a person typically sailing from the island at night and then deliberately overturning their canoe once out of sight of land. In 1981 I spent several weeks studying the geology of remote St Helena Island in the South Atlantic Ocean and heard stories about how, in times gone by, elderly people,

perceiving themselves as an intolerable burden on their families, would simply walk off the edge of one of the steep high cliffs that fringe this isolated island fortress.

26 Traditions about the periodic appearances of Burotu are recounted on pp. 163–8 of my 2009 book.

27 Details on pp. 104–5 of F. Morvan's 1980 book *Legends of the Sea*.

28 Quote is from p. 268 of K. Baarslag's 1944 book *Islands of Adventure*.

29 Weddell's considered view as to why he could not find the Auroras is that they never existed. He suggests that the Spanish surveyors aboard the *Atrevida*, tired and battling atrocious weather, mapped the Shag Rocks, some considerable distance away, which at the time had become attached to huge icebergs making the Spaniards think they must be ice-covered islands (from verbatim account in Stommel, 1984).

30 Quote from p. 89 of Volume 2 of Kerr (1824). Some equate Los Jardines with Enewetak Atoll (Marshall Islands) but this is located at 11°30'N, 162°20'E, not where Los Jardines was found.

31 From the 1940 account by Bryan in the *Proceedings of the United States Naval Institute*.

Chapter 6: Hidden Depths: Coastal Lands Submerged Long Ago

1 As suggested by Laura Botigué and others (2017, *Nature Communications*).

2 Much of the detail about Kongemose dogs in this paragraph comes from the work of Juliet Clutton-Brock and Nanna Noe-Nygaard (1990, *Journal of Archaeological Science*). Interestingly, their research shows that when living on the coast both Kongemose people and Kongemose dogs ate only the better-tasting sea fish rather than the blander freshwater fish living in the rivers and lakes. This evidence for fussy dog diets may resonate with you as it does with me.

3 This quote comes from p. 23 of Søren Sørenson's authoritative 2017 book about the Kongemose Culture.

4 See Vincent Riboulot and others (2018, *Nature Communications*).

5 This is from p. 235 of Ryan and Pitman's (1998) bestselling book about *Noah's Flood*.

6 Controversial archaeologist V. Gordon Childe named the shift from foraging to farming in Europe the Neolithic Revolution, implying an aggressive displacement of its earlier (Mesolithic) inhabitants by warlike scythe-wielding (Neolithic) farmers. Failing to find much evidence for conflict and indeed finding some (like interbreeding)

suggesting a more consensual process, many scientists today talk about the Neolithic Transition. Starting in the Near East some 10,000 years ago, the agricultural frontier moved like a wave west and north-west across Europe, reaching, for example, the Atlantic coast of Portugal some 7,600 years ago. The idea that the Neolithic Transition may have been driven by displacement of agriculturalists from the Black Sea Basin as a result of its (rapid) inundation was presented by Chris Turney and Heidi Brown (2007, *Quaternary Science Reviews*) but now appears untenable.

7   From p. 18 of the entertaining and solidly researched 2003 book *When the Great Abyss Opened* by J. David Pleins.

8   This research by Jens Herrle and others was reported in 2018 in *Scientific Reports*.

9   This event had huge consequences for global climate 8,400 years ago. May I also mention that modern Manitoba is strewn with lakes, 110,000 or so large enough to merit a name, tangible evidence of the former presence of Lake Agassiz.

10 This research was reported by Shi-Yong Yu and others in the 2007 issue of the journal *Geology*.

11 This basic model, which is central to Oppenheimer's 1998 book *Eden in the East*, is supported by linguistic and genetic data, although uncertainty remains owing to insufficient such data, especially in some parts of Sundaland.

12 In the key study by Pedro Soares and others (2008, *Molecular Biology and Evolution*), it was the study of mtDNA haplogroup E in contemporary populations of the emergent remnants of Sundaland that allow the convincing argument about population continuity in the area from at least 70,000 years ago, but also the conspicuous out-migration northwards about 12,000 years ago; 'the most plausible explanation for the spread of these genetic signatures is the impact on coastal-dwelling populations of the rapid global warming and sea level rises that led to the inundation of the Sunda shelf by meltwater at the end of the last ice age' (p. 1,215). An earlier study by Edwina Palmer (2007, *Japan Review*) argued something similar, namely that the Jomon peoples of the Japan archipelago could be shown through their material culture and languages to have originated in Sundaland, migrating northwards following its inundation.

13 The bottle-gourd, *Lagenaria siceraria*, was one of the earliest domesticated plants. It originated in tropical Africa and was dispersed by humans into Southeast Asia and thence across the Bering Strait into North America (N'dri, 2016, *Plant Molecular Biology Reporter*).

14 This comes from the 1937 book by Sarat Roy and Ramesh Roy
    entitled *The Kharias*.

15 Peter Buck's description of Hawaiki comes from his 1938 book,
    Vikings of the Sunrise, while the quotation in this sentence comes
    from p. 56 of Edward Tregear's 1891 *Maori-Polynesian Comparative
    Dictionary*.

16 The story in brief. In 1787, the HMS *Bounty* commanded by
    Lieutenant William Bligh was dispatched by the British Admiralty
    to Tahiti to gather breadfruit seedlings that could be planted on
    Caribbean islands to feed their growing population. By all accounts,
    Bligh was a harsh taskmaster, which prompted 12 crew members
    led by Fletcher Christian to mutiny. After setting Bligh and 19
    non-mutineers adrift in the Bounty's 7m launch on 28 April 1789,
    some mutineers settled on Tahiti. Fletcher and eight others settled
    with some Tahitians on Pitcairn, then uninhabited, and scuttled the
    Bounty; you can see its rudder today in the Fiji Museum. In the
    small launch, Bligh completed an astonishing 5,822km ocean
    journey from Tahiti to Timor and returned to Britain. In 1791, the
    HMS *Pandora* was sent to Tahiti where it captured 14 of the *Bounty*
    crew. Four drowned when the *Pandora* later struck a reef. Of the 10
    eventually brought back to Britain for trial, four were acquitted,
    three pardoned and three hanged. Of the mutineers who settled on
    Pitcairn, only one – John Adams – remained alive, a revered
    patriarch, when the island was visited by British warships in 1814
    and no punitive action was taken.

17 The pioneering work of Marshall Weisler on the human history
    of these remote islands shows that the people of Henderson were
    part of a mutually sustaining cross-ocean network with those of
    Pitcairn and Mangareva islands. From resource-poor Henderson,
    turtles and red feathers were traded for various items from Pitcairn
    and Mangareva including basalt for stone-tool manufacture, pearl
    shell for fish hooks, and even marriage partners.

18 South-East Polynesia is a discrete biogeographical unit with links to
    nearby island groups in French Polynesia and the Cook Islands
    where the flora derives from the Indo-Malesian region. To
    understand how astonishing this fact is, you could look at a globe
    or a map of the world and wonder how plants made their way
    independently from South Asia to the Pitcairn Island group, half a
    world of mostly ocean away. The likeliest explanation is that birds
    carried their seeds from island to island, although it has been
    proposed that coconuts colonised tropical South Pacific islands long
    before people did simply by bobbing along in the ocean currents

until they were washed ashore and could germinate. And do not discount the possibility of seeds reaching islands on floating rafts of vegetation or pumice, even the carcasses of dead whales.

19 Details of the flora of the Pitcairn group come from the research of Naomi Kingston and others (2003, *Journal of Biogeography*). Today all the plants living on low Ducie Atoll reached there through dispersal in water (hydrochory) while on Oeno Atoll, where there are habitats for birds, the figure is only 53 per cent. On higher Henderson Island, which is popular with seabirds of many kinds, most plants reached the island in this way (zoochory).

20 The most complete of Heyerdahl's pronouncements on the subject of the Easter Island biota is his 1989 book, optimistically (and misleadingly) entitled *Easter Island: The Mystery Solved*. The origin and the extraordinary antiquity of the Easter Island biota were explained by Newman and Foster (1983, *Bulletin of Marine Science*). Ideas about the timing of human arrival on Easter Island currently appear split between those who favour the longer-standing view that this was around AD 690 and those who infer a more recent date of around AD 1200.

21 Bleaching is a response by corals to stress, commonly overheating of ocean surface water. The coral polyps eject their symbiotic algae (which both feed them and give many their beautiful colours) resulting in the death of the corals and their loss of colour.

22 Many drowned reefs have been casualties of bursts of rapid sea level rise during post-glacial times. Imagine 8,000 years ago, a coral reef near the edge of the region within which ocean-surface temperatures were warm enough for corals to grow. The reef is surviving, but not doing as well as its counterparts in warmer waters. Then a meltwater dam on the nearest continent is breached, freshwater rushes out causing the sea level to rise over this reef perhaps 6.5m in 140 years. The corals simply cannot grow upwards fast enough at this location, so the reef becomes what has famously been termed 'a give-up reef'. It drowns.

23 I use alternative spellings to distinguish the island group (Hawaii) from the single island (Hawai'i). Use of the glottal stop in Pacific Island languages is very important. Think of it as replacing the letter k in a word.

24 A review of the causes of reef drowning around the Hawaiian Islands was given by Moore and Campbell (1987, *Science*). The deeply drowned Mahukona Reef was described by Clague and Moore (1991, *Bulletin of Volcanology*) and the drowning of the −150m reef discussed by Jody Webster and others (2004, *Geology*).

25   The anecdote comes from the 1983 book written by one of the
     scientists who made this discovery, Kenneth J. Hsu. The presence
     of an anomalously hard layer had actually been known earlier
     from seismic data but no one was then sure what it was.

Chapter 7: Deep in Shadow: Ancient Lands Now Hidden

1   Quotes from pp. 323 and 333 of Darwin (1870).
2   It takes a special kind of person to be an island biogeographer. I
    once picked up the late F. R Fosberg, one of the greatest island
    biogeographers, from an airport to transport him to a speaking
    engagement. He told me that in the 1930s he had visited the
    remote Pacific island of Vostok (now part of Kiribati), where
    much to his astonishment he could not find more than three
    species of terrestrial plant, a function of the island's comparatively
    small size and isolation. He wondered whether he should receive
    a prize for discovering the world's least diverse island flora.
3   Lister's 1891 account of 'Euan geology was published in the
    Quarterly Journal of the Geological Society of London; he wrote about
    how 'the whole eastern side, presented to the trade wind, is very
    abrupt, and rises in range above range of limestone-cliffs, the
    steep slopes between them being covered with dense wind-swept
    forest' (p. 598). The most comprehensive account of the geology
    of 'Eua was written by Edward Hoffmeister in 1932. On p. 23 of
    his memoir, published by the Bishop Museum in Hawaii, he
    became lyrical about the rare exposures of volcanic material,
    noting 'one of the most beautiful' is at Matalanga'a Maui in
    southern 'Eua. This is a natural limestone arch, said to be the
    place where once the mother of the demigod Maui, angered by
    the fact he had fallen asleep in the daytime when he was supposed
    to be weeding the food gardens, snatched his hoe to awaken him
    and threw it away where it made a massive hole in the rock from
    which Maui, sheepishly, later retrieved it.
4   This explanation is now widely accepted for the presence of the
    conifer genus *Acmopyle* on Viti Levu, the largest island in the Fiji
    group.
5   In a supreme irony, Burckhardt's identification of this ancient but
    vanished continent may have been influenced by the forerunners of
    the pseudoscience ideas that befog the whole subject for many
    people today. For in the nineteenth century, many European visitors
    to the Pacific Islands could not conceive how some, separated by

such vast distances of unbroken ocean, could be home to peoples with such similar cultural traits and languages, so they supposed that these islands were the mountain-tops of a huge continent that sunk. This idea was rejected by most geologists at the time, although Jules Garnier was a prominent exception in proposing a geological model that purported to support the anthropological scenario. Neither are correct, but both Garnier and Burckhardt were French and it is possible that Burckhardt had read Garnier's (1870) book outlining his lost-continent theory before he began studying the turbidites of western South America.

6   In his 1922 article in *The Geographical Journal*, J. M. Wordie argued that a Viking landing from Iceland on Jan Mayen was more plausible than one on the more distant Svalbard (Spitsbergen).

7   '... et passant à travers des bandes d'oiseaux de mer, poussant leur curiosité si loin, qu'on est obligé de gesticuler pour les écarter un peu', from p. 10 of the 1891 book, *Voyage de La Manche A L'Ile Jan-Mayen*.

8   Aside from observing that the ocean floor below/around Jan Mayen was unusually shallow, suggesting it was made from less-dense material than the oceanic crust, geologists concluded that Jan Mayen was indeed a sliver of continental crust in two ways. First, from looking at samples of the rocks being brought up by drilling, it was found these were different not just in composition, but also far older than the rocks that usually compose the ocean floor. The conclusive evidence that there was a microcontinent below Jan Mayen came through measuring the gravity, magnetic and seismic characteristics of this part of the ocean floor. Microcontinents are usually places where there are positive gravity anomalies, meaning that the Earth's crust below the surface is lighter (less dense) than expected. In addition, they are magnetically quiet, meaning that they do not exhibit the young palaeomagnetic signatures that are characteristic of oceanic crust. Finally, by sending artificially generated seismic (earthquake-like) waves through ocean-floor rocks and noting their travel times back to the Earth's surface, the density and heat characteristics of these rocks can be determined. In this way, the Jan Mayen microcontinent was demonstrated to have a continental root extending 15km beneath the ocean floor and having seismic velocities of 6.7–7km/second, typical of continental crust. Case closed.

9   On p. 243 of his 1874 book on *The Structure and Distribution of Coral Reefs*, Darwin states that he 'took pains to procure plans and information regarding the several islands' of the Seychelles from Captain F. Moresby and one Dr Allan of Forres.

10 In contrast to its northern continental part, the southern part of
    the Mascarene Plateau is wholly oceanic in origin, the product of
    volcanism at the Réunion hotspot over the past 35 million years
    or so.

11 The microcontinent at the northern end of the Mascarene Plateau
    has been subject to extensive geological investigation in recent
    decades. Seismic studies show a crustal thickness of some 33km –
    far greater than would be expected for oceanic crust. The presence
    of thinner oceanic crust all around confirmed the prescient belief of
    Alfred Wegener, generally hailed as the originator of the theory of
    continental drift, that the Seychelles were Gondwanan, a
    microcontinent rifted off the Indian subcontinent.

12 The issue of whether most LIPs originate from mantle plumes is
    controversial. Hoping to resolve the question, The Great Plume
    Debate was held in 2005 where both sides argued their points. It
    was concluded that the Debate 'failed to land a telling blow on
    the mantle plume chin, and no creditable alternative emerged to
    explain the principal features of LIPs', quoted on p. 268 of
    Campbell's 2005 article in *Elements*.

13 An impact origin for the Ontong Java Plateau was proposed by
    Ingle and Coffin (2004, *Earth and Planetary Science Letters*).

14 All quotes from June Young (née McLauchlan) are in a 2007
    compilation of New Zealand disaster stories (GNS Science
    Report 2007/05).

15 Quotes from p. 341 of Eiby (1982, *Journal of the Royal Society of
    New Zealand*).

16 Details from R. Bell and others (2014, *Earth and Planetary Science
    Letters*). It is my interpretation that the 1947 tsunamis were caused
    by a submarine landslide, based on my reading of Jonas Ruh's
    2016 article in *Terra Nova*.

Chapter 8: Earth's Watery Shroud: Sea Level Changes

1 The closing lines of the 1851 novel *Moby Dick* by Herman Melville.

2 I made a similar point in the 2020 issue of the *Chicago Quarterly
   Review*, writing that 'science could have learned more and sooner
   had it treated the "stories" of ancient Australians (and people of
   other non-western cultures) more seriously as potential sources of
   information about the past, had it not pejoratively dismissed such
   stories as entertainments rather than expository, characterized the

storytellers as "literati" rather than true scientists communicating their wisdom along unfamiliar pathways'.

3 David Thompson was named Koo-koo-sint (stargazer) by some native Americans. In 1820, his essence was captured by a dinner companion – 'Never mind his Bunyan-like face and cropped hair; he has a very powerful mind, and a singular faculty of picture-making. He can create a wilderness and people it … or climb the Rocky Mountains with you in a snow-storm, so clearly and palpably, that only shut your eyes and you hear the crack of the rifle, or feel the snow-flakes melt on your cheeks as he talks' (Bigsby, 1850, *The Shoe and Canoe*, pp. 113–4).

4 From the *Journal of David Thompson*, copied from the original in the Toronto archives by T. C. Elliott and published in 1911.

5 Adapted from p. 2 of the US government's 1974 booklet about *The Channeled Scablands of Eastern Washington*, author unknown.

6 James Hutton is described as 'a caricature of the dour Scot' on p. 176 of K. S. Thomson's well-informed *Before Darwin*.

7 Almost a century ago, with characteristic brio, the New Zealand ethnologist Elsdon Best captured this point thus. To him, the first Pacific Islander 'was the champion explorer of unknown seas of neolithic times. For, look you, for long centuries the Asiatic tethered his ships to his continent ere he gained courage to take advantage of the six months' steady wind across the Indian Ocean; the Carthaginian crept cautiously down the West African coasts, tying his vessel to a tree each night lest he should go to sleep and lose her; your European got nervous when the coast-line became dim, and Columbus felt his way across the Western Ocean while his half-crazed crew whined to their gods to keep them from falling over the edge of the world' (1923, *Polynesian Voyaging*, p. 9).

8 Much of the information in this section comes from Mike Carson's comprehensive 2018 book, *Archaeology of Pacific Oceania*, on p. 125 of which he writes that the earliest decorated pottery, dating from about 3,500 years ago, found on at least three islands in the Mariana group in Micronesia 'match exactly' with the pottery being made around this time in the northern and central Philippines.

9 Quotes from the article by M.G. Vilar and others (2013, *American Journal of Human Biology*).

10 The conclusion proved inescapable to linguist Robert Blust (2000, *Oceanic Linguistics*).

11   There is a possibility that people in this part of the Philippines at this time were also cultivating rice and millet, neither of which made it to the Marianas with the first people.

12   This argument was first put by me in my 1999 book on *Environmental Change in the Pacific Basin* yet I fear I may also be culpable of what my friend Mike Carson considers wild speculation 'about the motivations of ancient sea-going migrants' (2018, p. 175).

13   If this is of particular interest to you, the 2018 collection of studies on *Ancient Psychoactive Substances* edited by Scott Fitzpatrick is an excellent source of further information. It is possible that the crimson mouths of habitual betel nut chewers were mistaken by excitable foreigners, encountering Pacific Island people for the first time, as evidence for the habitual consumption of human flesh, giving rise to exaggerated claims about cannibalism in these islands. When Captain Cook landed on Niue Island (South Pacific) on 2 June 1774, he encountered islanders who had painted their faces with the juice of red bananas. Hasty to judge, he wrote: 'The Conduct and aspect of these Islanders occasioned my giving it the Name of Savage Island.' A damning and ill-informed judgement that lasted centuries.

14   The study of ancient diets here was by García Guixé and others (2006, *Current Anthropology*). The most common shellfish consumed by the Mesolithic coastal peoples of eastern Spain 9,500 years ago was the saltwater clam, *Cerastoderma glaucum*, still consumed regularly here today.

15   Most of the detail here comes from the study by Elodie Brisset and others (2018, *Global and Planetary Change*).

16   Carl O'Brien's article in the *Irish Times* (25 March 2009) refers to research which found that Ireland is 'a nation of brooding pessimists'.

17   This quotation refers to the French and is based on research by Claudia Senik who found that the French 'culture and mentality' make them far less happy than you would expect given their wealth and lifestyle (2014, *Journal of Economic Behavior and Organization*).

18   I intend to be provocative here. But I am convinced that 'sunken land' and 'sunken city' stories are far more numerous along the Atlantic coasts of north-west Europe than elsewhere simply because of this region's anomalous history of post-glacial sea level rise (more in Nunn, 2020, *Environmental Humanities*).

19   This is the idea of Mark Clendon (2006, *Current Anthropology*) who argues that the Pama-Nyungan language group spread across

Australia as the land bridge connecting it to Papua New Guinea was flooded after the last ice age ended and its speakers were forced to disperse.

20   The date of 11,700 years ago for the intrusion of seawater into freshwater Lake Carpentaria comes from research by Craig Sloss and others (2018, *The Holocene*). Brackish water conditions continued in the former lake until about 10,000 years ago when it became fully connected to the ocean. Sea level rise continued around this part of the Australian coast, present sea level being reached about 7,700 years ago.

21   This quote is from p. 20 of Dick Roughsey's 1971 autobiography, *Moon and Rainbow*.

22   In the context of Aboriginal Australia, there are a number of coastal drowning stories of similar antiquity (Nunn, 2018).

23   The relevant 1927 study by archaeologist Osbert Crawford was published in the journal *Antiquity*. His prescience is also demonstrated by this passage, relevant to this book; 'It is a common mistake to suppose that an 'uneducated' person is less intelligent or less accurate in observation than one who has acquired book-knowledge' (p. 12). I know a few scientists might bristle at such a suggestion.

24   Gary Robinson (2013, *Internet Archaeology*) notes during the period 5,000–3,500 years ago, 'the modern islands of St Mary's, Bryher, Tresco and St Martins formed one large island' (part 4). The evidence for Neolithic occupation of the Scillies is plentiful.

25   The first quote comes from a translation of the *Chronicle of Florence of Worcester*, completed in 1140 (Hunt, 1903, *Popular Romances*), the second from *Britannia* (Camden, 1590).

Chapter 9: 'The Island Tilts … The Tourists Go Mad': The Ups and Downs of the Earth's Crust

1   A humorous aside is that the local people organising this trip were accustomed to taking tourists to one of the fifteen familiar volcanic islets but the one on which they had been requested to land was 'an extra one, and they had better have tea on one of the ordinaries' which they did not. Although Forster's island is fictitious, he probably based this part of his story on the short-lived existence of Isola Ferdinandea (Graham Island) that appeared for a few months in 1831–32 south-west of Sicily and

received several intrepid visitors; towards the end of its life, it had fractured into 12 pieces. More on this and similar islands in Chapter 11.

2   The first quote is from a 1692 broadsheet by one Captain Crocker entitled 'A True and Perfect Relation of that most Sad and Terrible Earthquake, at Port-Royal in Jamaica'. The second quote comes from the 2000 History Today article by Larry Gragg. I recommend the book, Apocalypse 1682, by Ben Hughes for a lively account of the 1692 Port Royal earthquake.

3   Both extracts in this paragraph are found in the account of Gragg, mentioned above. The Reverend Heath's hopes were not realised; within a few years, Port Royal had regained its former reputation.

4   The Enlightenment was an important stage in western thinking that saw faith-based understanding replaced by empirical-based understanding. Implicit in this was the premise that the world, while possibly created by God, was nevertheless understandable and could therefore be manipulated for the benefit of humanity. The Enlightenment is normally regarded as an eighteenth-century phenomenon.

5   You do not need to look very hard to find examples. In January 2005, a month after the Indian Ocean tsunami in which about 230,000 people died, thousands of people marched in Morocco in support of the view that the tsunami had been an expression of God's displeasure with the sex tourism industry of Southeast Asia. In the same year, after Hurricane Katrina devastated New Orleans, one religious commentator, using reasoning astonishingly similar to that aired after the 1692 Port Royal earthquake, claimed this had happened because pre-storm New Orleans had 'a level of sin offensive to God'. In January 2010, following the earthquake in Haiti in which some 250,000 people perished, a conservative 'televangelist' claimed this was punishment for Haitians' eighteenth-century 'pact with the devil'. None of these explanations is evidence-based.

6   From an online compilation of eyewitness accounts. Another is as follows. 'An eyewitness who was in a service station on the west side of the highway said he and a companion ran out the east door of the building as the concrete floor began to crack. They got about three feet out of the building when a crack about three feet wide opened between them. He said that cracks formed about each of them, leaving each man on a small island about three feet wide that moved up and down. He said it was like riding an open

elevator. As he went down, the other man went up. And then they'd pass each other going in opposite directions. He said that the earth all around them broke into similar pieces and that as the blocks of frozen earth moved up and down, the cracks also opened and closed causing muddy water to spout as high as 50 feet. He said after the shaking stopped, water filled the open cracks. He estimated the duration of the quake at about four to five minutes.' (From *Chronology of Physical Events of the Alaskan Earthquake*, 1966, University of Alaska Anchorage.)

7   The whole story is well told in Henry Fountain's 2017 book, *The Great Quake*.

8   In 2018, Kesavan Veluthat, Director of the Institute for Heritage Studies of Coastal Kerala, quipped: 'I often joke that Kerala isn't a part of India, it's a part of the Indian Ocean. It was the Western influence that shaped it,' (*The Hindu*, 29 September 2018).

9   The inference that the author of this *Periplus* was not overly educated is from p. 16 of Schoff's 1912 edition of the journal.

10   Quoted in translation by Anitta Kunnappily (2018, *International Journal of Maritime History*).

11   This evocative description is from an article in *The Guardian* by Srinath Perur on 10 August 2016.

12   The report of a 'severe earthquake' in AD 1341 on the Malabar coast of India is in Ballore's 1900 review in the *Memoirs of the Geological Survey of India*. My contention that Vypin Island was raised during this earthquake rather than by flood is rejected by C.P. Rajendran and others (2009, *Journal of the Geological Society of India*), but I find their argument a little timid given that they conclude that 'the central midland district [where Vypin lies] is more prone to seismic activity compared to other parts of Kerala' (p. 785). I suspect the orthodox view that Vypin formed as a result of a flood rather than an earthquake originated with an 1846 English account of the island which reported it was 'thrown up by the sea about the year 1341' (Newbold, 1846, *Journal of the Royal Asiatic Society of Great Britain and Ireland*, p. 252), an ambiguous use of language which was subsequently misinterpreted to mean 'thrown up' by a flood not an earthquake. Of course, I could be completely wrong.

13   This quote is on p. 317 of Volume 1 of Alexander Hamilton's 1739 *A New Account of the East-Indies*. Later, in 1772, it was reported of Calicut that 'several vessels have been wrecked upon the ruins of the old city, now under water ... Every vestige of that magnificent city is now whelmed beneath the sea ... at very low

water I have occasionally seen the waves breaking over the tops of the highest temples and minarets, but in general nothing is to be distinguished of the ancient emporium' (Forbes, 1813, *Oriental Memoirs*, Volume 1, pp. 322–3).

14   There is no way that 'a sudden rising of the sea' can cause land to be permanently submerged unless the land itself sank at the same time. The quote (updated English) comes from p. 184 of Philip Dormer Stanhope's 1785 collected letters. Currently unpublished research by myself and Roselyn Kumar suggests that Da Gama initially visited a place he thought was Calicut, but which was not. Only on a later visit to the Kerala coast did he discover the real Calicut, which he bombarded and which was later completely destroyed by a combination of subsidence and tsunamis.

15   This comes from p. 796 of the thoughtful article by C. P. Rajendran and others (2009, *Journal of the Geological Society of India*).

16   Many Pacific examples are described in my 2009 book.

17   In 2006 with three colleagues, I wrote an article in the *Journal of the Royal Society of New Zealand* reporting our research into the disappearance of Mamata; the full name, Vanua Mamata, likely means 'dry or shallow (low-lying) land'.

18   This is a condensed version of the story told in 1914 in the Vao language of Vanuatu, translated and written by John Layard (1942).

19   This description of Cone (Ritter) Island comes from p. 197 of Jacobs (1844, *Scenes*).

20   Details here from Karstens and others (2019, *Earth and Planetary Science Letters*). The silhouette of modern Ritter Island in Figure 9.3 comes from research by Eli Silver and others (2005, *Eos*).

21   The first comprehensive study of the 1888 Ritter Island collapse was by Steven Ward and Simon Day (2003, *Geophysical Journal International*) who also gained international attention for their modelling of a possible future collapse of some islands in the Canary Islands (2001, *Geophysical Research Letters*), discussed in Chapter 13.

Chapter 10: Falling Apart: Collapsing Continents and Islands

1   Much of the research about the Störegga Slide has been carried out by Stein Bondevik and his collaborators; the moss (*Hylocomium splendens*) research was reported in 2015 by Rydgren and Bondevik in the journal *Geology*.

2 The deduction that the Störegga Slide occurred in autumn was
   confirmed by Sue Dawson and David Smith (2000, *Marine Geology*)
   who found wild cherry stones in Störegga tsunami deposits along
   the northern Sutherland coast of Scotland; these trees (*Prunus
   avium*) commonly fruit in late September and October.

3 See Ian Boomer and others (2007, *The Holocene*).

4 First suggested by David Smith, this idea was later written up by
   Weniger and others (2008, *Documenta Praehistorica*) and is central
   to the work of Vince Gaffney and others (2009).

5 See Bondevik and others (2012, *Quaternary Science Reviews*).

6 Ages here come from the study by Törnqvist and Hijma (2012,
   Nature Geoscience).

7 From p. 15 of the book about Pinçon by J. R. McClymont (1916).

8 Others in Pinçon's crew, perhaps informed by local people, named
   the river El Rio Marañón. The name Amazon is European, given to
   the area (Amazonas) and thence to the river following a report by
   Francisco de Orellana of a battle he had fought on its banks with
   both men and women of the Tapuyas tribe. The existence of warrior
   women put Orellana in mind of those of North Africa whom the
   Roman traveller Herodotus had named Amazons.

9 Herrera, quoted by McClymont (1916).

10 Each of these individual slides is approximately 15,000km2 in
   area (the size of Jamaica), has a maximum thickness of 200m, and
   contains some 5,000 gigatonnes of sediment. The research
   discussed here was reported by Mark Maslin and others (2005,
   Quaternary Science Reviews).

11 Gas hydrates make up 10 per cent by volume of some of the
   landslide bodies within the Amazon Fan, demonstrating that the
   part of the Amazon Delta which collapsed was rich in gas hydrates.
   It might therefore seem an obvious inference to assume that
   dissociation (break-up) of these gas hydrates was the principal cause
   of collapse. Not so fast! It was suggested by Maslin and collaborators
   (see previous note) that the gas hydrates were not the cause of
   collapse but an effect of it. Imagine the edge of the Amazon Delta
   suddenly being sliced off; the ice-encased gas hydrates deep within
   the sediment pile would suddenly be liberated (but not dissociated),
   incorporated in free form into the sediments that tumbled
   downslope along the sea floor to their ultimate resting place in the
   Amazon Fan.

12 The orthodox view is that South America was reached by people
   only after they had reached North America, the first migrants
   crossing the Bering Strait after the end of the last ice age about

14,000 years ago when it was (mostly) dry land. The more unbending exponents of this view have long been troubled by the possibility of far earlier ages for people in South America than the North. An example of these comes from the Toca do Boqueirao rock shelter in the Serra da Capivara National Park where charcoal in an old campsite (hearth) suggests people were preparing pigments here to create the stunning rock art 24,000 years ago (Lahaye and others, 2015, *Quaternary Geochronology*). A resolution of this issue may be that the earliest Americans paddled along the continents' western edge, occupying islands like Haida Gwaii (discussed in Chapter 8) and reaching South America, long before they made significant inroads into the North American continent.

13   The phrase and most of the research cited here comes from research by Chris Goldfinger and others (2000, *Pure and Applied Geophysics*).

14   Depending on the season, the Sahara has expanded anywhere between 11 per cent and 18 per cent within the past hundred years or so, largely in response to climate change (Thomas, 2018, *Journal of Climate*).

15   The research described here is that of R. Henrich and others (2008, *Marine and Petroleum Geology*).

16   See research by Jessie Creamean and others (2013, *Science*).

17   Translated from the Spanish, Englert's most comprehensive account of Easter Island traditions is the posthumously published *Island at the Center of the World* (1972).

18   Many such stories are related in my 2009 book, *Vanished Islands and Hidden Continents of the Pacific*. It seems likely that people first reached Easter Island at least 700 years ago, meaning that the story of Uoke must be at least that old.

19   The original paper trail was uncovered and presented by Paul Filmer and colleagues (1994, *Marine Geophysical Research*).

20   Hergest's tilt at immortality was to name the northern group of the Marquesas the Hergest Islands. The name didn't stick.

21   The story is on pp. 67–68 of the book *Islanders of the Pacific* by T. R. St. Johnston (1921).

22   This work was described by Hélène Hébert and others (2001, *Journal of Geophysical Research*).

23   Movement within the Hilina Slump on the south-east flank of active Kilauea volcano on Hawai'i Island in November 2000 was not earthquake-triggered but followed nine days of incessant rain that lubricated slip planes. An array of Global Positioning System (GPS) receivers was then set up across the Hilina Slump over a

36-hour period and showed a 20km long and 10km wide piece of the Kilauea volcano to be moving seawards at a rate of around 5 cm per day (Cervelli and others, 2002, *Nature*).

24 Most details about the 1999 Fatu Hiva landslide and tsunami were reported by Emile Okal and others (2002, *Bulletin de la Société Géologique de France*).

25 The work on Hierro was reported by D. G. Masson (1996, *Geology*).

26 This is the process which best explains the present situation at the Cumbre Vieja volcano, the 1949 movement of which appears to have led to a concentration of magma upwelling along the line of weakness created (Day and others, 1999, *Journal of Volcanology and Geothermal Research*).

27 For his uncompromising nature, Harold Stearns's wife, Claudia, nicknamed him 'thistle-tongue'.

28 Along the wind-blasted arid south coast of Lana'i Island, Anne Felton undertook a meticulous analysis of the Hulopoe Gravels to reach this conclusion (2000, *Pure and Applied Geophysics*).

29 The bump in the ocean floor is that caused as a result of the enormous island of Hawai'i (the Big Island) pressing down on the surrounding Earth's crust and deforming it, producing a moat and an arch. The arch is the bump, rather like the situation shown in Figure 8.2A. The key work is that by Barbara Keating and Chuck Helsley (2002, *Sedimentary Geology*).

30 The research was reported by Jody Webster and others (2007, *International Journal of Earth Sciences*) and is based on studies of two submerged shorelines (150m and 230m below sea level) off Lana'i Island above which water depth has not changed much for 30,000 years.

Chapter 11: 'Huge and Mighty Hilles of Water': Monstrous Waves

1 From a pamphlet published shortly after this event entitled *God's Warning to his People of England*, quoted on p. 131 of the 1951 book about *Monmouthshire* by Arthur Mee.

2 The quotation comes from p. 78 of the 1930 reprint of La Pérouse's account of his round-the-world voyage. For a naval commander trained to resist embellishment, La Pérouse becomes strangely lyrical when describing Lituya Bay, noting that it is 'bordered by sheer mountains, extremely high, snow-covered, without a blade of grass on this immense pile of rocks, condemned by Nature to eternal sterility. I never saw a puff of wind cause a ripple on the

water surface; it is disturbed only by the fall of enormous chunks of ice that are frequently detached from five different glaciers, and which can make, in falling, a noise that resounds through the mountains. The air is so calm and the silence so profound, that the simple voice of a man can be heard for half a mile, even over the noise of the many seabirds that lay their eggs in the rock crevices.' In the original, '*un bassin d'eau bordé par des montagnes à pic, d'une hauteur excessive, couvertes de neige, sans un brin d'herbe sur cet amas immense de rochers condamnés par la nature à une stérilité èternelle. Je n'ai jamais vu un souffle de vent rider la surface de cette eau: elle n'est troublée que par la chute d'énormes morceaux de glace qui se détachent très fréquemment de cinq différents glaciers, et qui font, en tombant, un bruit qui retentit au loin dans les montagnes. L'air y est si tranquille et le silence si profond, que la simple voix d'un homme se fait entendre à une demilieue, ainsi que le bruit de quelques oiseaux de mer qui déposent leurs œufs dans le creux de ces rochers.'*

3   The most comprehensive account of the giant-wave history of Lituya Bay is that by Miller (1960, *United States Geological Survey Professional Paper 354-C*).

4   Quote is from p. 40 of Roberts (2005, *History Today*), from which other incidental details are taken.

5   As often happens in such instances, these islands grew back after 15 years or so (Andrade and others, 2004, *Marine Geology*).

6   From p. 397 of the 1839 book, *Tales About the United States*, by Peter Parley.

7   Pohnpei is one of my favourite places but, if you ever plan to visit, pack a sturdy umbrella. Owing to the steepness and height of the volcano that forms Pohnpei Island, its highest peaks receive more than 8m of rain each year.

8   Details are from pp. 81–83 of the 1988 book *Upon a Stone Altar* by David Hanlon. The battle of Nahlapenlohd was arranged as a deciding context between the warring clans of Kitti and Madolenihmw. The Kitti forces were supported by a Londoner named James Headley and a Filipino named Narcissus de Los Santos, who operated a small cannon that fired sections of anchor cable (as ammunition) into the Madolenihmw warriors with great effect.

9   Most of this research was reported by myself, Gus Kohler and Roselyn Kumar in the 2017 issue of the *Journal of Coastal Conservation*.

10 A key study is that by Arthur Webb and Paul Kench (2010, *Global and Planetary Change*). The apparent implication of this research – that rising sea level is not causing small fragile islands to

disappear – made headlines around the world but is, I suggest, a misleading inference. The central problem is that these researchers assess changes of these islands (which are little more than piles of sand and gravel) in two dimensions whereas they are of course three-dimensional. So, apparent extension of the shoreline in one place could simply be a result of redistribution of sediments within the island pile rather than representing actual growth.

11   Research reported by Martin Arntsen and colleagues in a 2019 issue of the *Journal of Geophysical Research – Oceans*.

12   In most polar and sub-polar regions, the existence of frozen groundwater within the soil and surface sediments gave rise to a permanently frozen layer (permafrost). As hard as concrete, many buildings and infrastructure in these parts of the world have their foundations within permafrost. Thawing permafrost not only destabilises these, but also releases bacteria and viruses, to some of which humans have not been exposed for thousands of years.

13   Quoted in *The Barents Observer* on 22 February 2019.

14   Historical detail about Billingsgate Island is found in the 1991 book, *A History of Billingsgate,* by historian Durand Echeverria.

15   Actually 4.3 x 1018 J, equal to a 100-gigaton bomb. This and other details in this section come from the insights of Roger Bilham (2005, *Science* and elsewhere).

16   Quoted on p. 76 of the 1999 book entitled *Vulnerability and Adaptation to Climate Change for Bangladesh* edited by Saleemul Huq and others.

17   This refers to the period 1973–2010 and comes from research by Md Shahjahan Ali and colleagues (2013, *Dhaka University Journal of Biological Sciences*).

18   The information was reported to the New Zealand authorities and was obtained from the Department of Scientific and Industrial Research (Geophysics) in Wellington.

19   The most comprehensive reference on Myojin-sho is the work by Kokichi Iizasa and others (1999, *Science*). Some of the other volcanoes growing up from the rim of the Myojin Knoll caldera are the Bayonnaise Rocks and Fukutokuoka-no-ba. The latter formed sizeable islands on at least four occasions during the twentieth century. In January 1986, an eruption of Fukutokuoka-no-ba produced a 10m-high island measuring 600m by 200m that disappeared within three months.

20   It is fairly certain that this is what happened, but there were no survivors to tell the tale. So, our certainty comes from understanding how degassing of an underwater volcano can

lower the overlying ocean-water density, causing ships to sink. It also comes from stories of near-misses, like that of the steamer *Kilauea* which in late February 1877 entered Kealakekua Bay on the island of Hawai'i to find columns of steam and volcanic gas rising upward from the water as well as blocks of semi-molten lava floating on the ocean surface – a result of the only submarine eruption recorded in the island's history. The captain of the *Kilauea* steered away to a safe distance.

21 Local Sicilians speculated that Ferdinandea disappeared as quickly as it could because it resented being a British colony in an Italian sea!

22 Baker first arrived in Tonga in 1860 as a Wesleyan missionary, but soon became embroiled in national politics, much to the annoyance of the other missionaries. Following his deportation, after some time in Australia, Baker returned to Tonga in 1890, gave up his Wesleyan allegiance, became Premier of Tonga, and founded the Free Church – which all Tongans were then required to join.

23 Quote from pp. 41–43 of the 1886 article by Baker in the *Transactions of the New Zealand Institute*.

24 More details about the eruptive history of Fonuafo'ou and other underwater volcanoes in Tonga can be found in my 1998 book, *Pacific Island Landscapes*. The role of shallow underwater eruptions in Pacific Islander myths is described in my 2009 book, *Vanished Islands and Hidden Continents of the Pacific*.

25 See the 2007 article by Corey Howell in *Melanesian Geographic*. Howell has seen fish affected by subsonic 'whumps' at Borokua Island, 40 nautical miles east of Kavachi.

26 Kavachi has erupted at least nine times since 1939 (Nunn, 1994).

Chapter 12: Volcanic Islands

1 *Insula hec in Anno Dñi 1456 fuit totaliter cōbusta* in the original.

2 Echoing my view, albeit more tortuously, geographer William Babcock noted that although there is no available evidence the unnamed island mapped by Ruysch existed, 'the assertion is not in itself incredible' (1922, p. 175).

3 From p. 33 of Peter Hallberg's (1975) *Old Icelandic Poetry*.

4 Most of the information about Minoan Akrotiri on Stronghyle/Santorini/Thera comes from the beautifully illustrated book, *Akrotiri*, by Clairy Palyvou (2005). Not to be confused with modern Akrotiri on the island of Crete.

5 The precise age quoted for the caldera-forming eruption comes from Walter Friedrich and others (2006, *Science*) who were

fortunate to discover a Minoan-era olive tree buried by pumice and ashes from this event. They used radiocarbon dating to determine the age of tree-ring segments, bracketing the eruption date to 1627–1600 BC.

6   Recent research by Irini Papageorgiou on an Akrotiri fresco shows that a detail, long interpreted as a dovecote, is more likely an aviary, suggesting that the Minoan inhabitants of Stronghyle kept bees and consumed and perhaps even traded honey. Another Akrotiri fresco depicts women gathering saffron, valued as a cloth dye at that time.

7   Most geological details are from Paraskevi Nomikou and others (2016, *Nature Communications*).

8   This quote comes from p. 28 of the 1998 book *Imagining Atlantis* by Richard Ellis, a bold attempt to stand up against the tidal wave of misinformation about Atlantis threatening to engulf rational thought at that time.

9   J. R. R. Tolkien, renowned author of *The Lord of the Rings*, had a recurrent obsession with Atlantis myths, often dreaming of 'the stupendous and ineluctable wave advancing from the Sea or over the land, sometimes dark, sometimes green and sunlit', as he wrote in a letter to W. H. Auden (Carpenter, 1981, *The Letters*, p. 361). In *The Lord of the Rings*, Tolkien's recurring dream of the great wave that spelled the end of Númenor haunts the character Faramir of Gondor.

10 English translation by Christy Haruel of a story *The Legend of Kuwae* told in Bislama by Chief Tom Tipoamata of Tongoa Island in Vanuatu.

11 From pp. 13–16 of Michelsen's 1893 book about *Missionary Perils and Triumphs in Tongoa*. Michelsen also recorded that local traditions attribute the absence of snakes on Tongoa and other islands that were once part of Kuwae to 'the complete desolation wrought by the catastrophe' (p. 15).

12 The original French term comes from the initial report by J-P Eissen and colleagues (1994, *La Recherche*), while the detail that the Kuwae eruption was one of the largest in the last 10,000 years comes from the expanded work by Michel Monzier and others (1994, *Journal of Volcanology and Geothermal Research*).

13 In central Mexico, the 1454 *Famine of One Rabbit* is attributed to the Kuwae eruption. Graphic descriptions of the devastating consequences of this famine based on accounts in Aztec codices are in the article by Matthew Therrell and others (2004, *Bulletin of the American Meteorological Society*).

14 Most details quoted about the fifteenth-century siege of Constantinople come from the compelling 2005 book, *Constantinople: The Last Great Siege 1453*, by Roger Crowley.

15 A summary of ice-core evidence for the eruption of Kuwae is given by Chaochao Gao and others (2006, *Journal of Geophysical Research – Atmospheres*).

16 '... in Vanuatu natural disasters are perceived as social rather than natural events' wrote Jean-Christophe Galipaud in an insightful chapter in a 2002 volume on *Natural Disasters and Cultural Change*.

17 The spelling Krakatoa (rather than Krakatau) may actually have been a typographical error introduced to the global English lexicon only after the 1883 eruption. This and other gems are in the entertaining (yet paradoxically titled) book *Krakatoa: The Day the World Exploded* (2003) by Simon Winchester. I cannot resist drawing attention to the title of the 1969 Hollywood film entitled *Krakatoa, East of Java* starring Maximilian Schell. Krakatau/Krakatoa is actually west of Java!

18 Vatukoula means 'rock of gold' in Fijian (*vatu* = rock; *koula* = gold), and you might deduce from this that an inordinate love of gold was something innate in millennia-old Fijian culture and indeed every human culture – but that would be wrong. In Fiji, before Europeans arrived in the nineteenth century and introduced the idea of gold as the most precious colour and substance, Fijians (and many other Pacific islanders) considered red, particularly red feathers (*kula*) of the endemic lory (*Phigys solitarius*), to represent the epitome of wealth and power. And of course, the indigenous Māori of New Zealand appear to have treated green in a similar way, creating fabulous carvings today and in the past, from the naturally occurring greenstones (*pounamu*) in these islands. The ancestors of the Māori, who were the same as the earliest Fijians, may also have regarded green in this way. With Sepeti Matararaba and Roselyn Kumar, I spent six years excavating the extraordinary 3,000-year old settlement site at Bourewa in Fiji where, in addition to intricately decorated pieces of pottery and exquisitely carved shell jewellery, we found hundreds of stone tools manufactured from dacite, a fine-grained (thus unusually hard-wearing) rock with a stunning green colour.

19 This detail is from the article by Iizasa and others (1999, *Science*). Not surprisingly, such Kuroko-type deposits have attracted considerable attention from mineral exploration companies in the past few decades.

20 This proposition is the subject of a serious and compelling scientific study (Olson and others, 2007, *Environmental History*). The association of livid 'Krakatau sunsets' with distant fire led to despondency, even panic, in some parts of the world. In Honolulu, the Reverend Sereno Bishop thought these unusual sunsets had a 'peculiar lurid glow, as of a distant conflagration' while in New York and Connecticut local fire brigades were summoned by alarmed residents. The imagery also influenced contemporary poets. For example, Tennyson, imagining in 1892 the evening cogitations of St Telemachus, had him inspired by sunsets of 'lurid crimson' that led him to ask:

> ... *Is earth*
> *On fire to the West? or is the Demon-god*
> *Wroth at his fall ...*

21 In this regard, we should note that the 1883 Krakatau eruption was perhaps the first disaster of which details were relayed rapidly around the Earth thanks to the recently installed ocean-floor telegraph cable. For this reason, news of the Krakatau eruption was received four hours after it happened on the other side of the world in Boston. Compare that with the news of President Lincoln's assassination a mere seventeen years earlier, before the telegraph existed, which took twelve days to reach the rest of the world.

22 Please note that none of this is true, merely a product of Churchward's excitable imagination. The quotation comes from p. 15 of his 1931 book *The Children of Mu*.

23 Blavatsky's influence on modern new-age and pseudoscience thinking is significant, and it is possible to trace many current new-age and millennialist ideas back to *The Secret Doctrine* (whence the quotes in this section come) and thence back to the Krakatau eruption.

24 This suggestion was first made by Dr Jan Lindsay.

Chapter 13: Slipping into the Shadows: Vanishing Lands

1 Finishing the final draft of this book in March 2020 during the coronavirus pandemic, these words took on new significance.

2 Even though the Japanese government now appears to be having second thoughts about investing millions in the pointless repositioning of the Nadi River mouth, its money would be far better spent funding the infrastructural skeleton of a new Nadi

Town on higher ground at the rear of its present site (see my 2019 article with Roselyn Kumar in the inaugural issue of *One Earth*).

3   An excellent book chapter reviewing such 'nature-based solutions' was published in 2019 by Virginie Duvat and Alexandre Magnan and is free to download at https://tinyurl.com/vjhguaj

4   In his 1989 book *Global Warming*, pioneering climate-change scientist Stephen Schneider devoted a chapter (humorously entitled 'Mediarology') to a discussion of the way that the media often unhelpfully reports the climate-change issue by emphasising extreme projections and unrealistic scenarios simply because they are newsworthy.

5   The IPCC assessment report referred to in this section is the 5th Assessment Report (5AR) that was issued in 2013–14. I was a lead author for the chapter on sea level change in this. My view that the 6th Assessment Report (6AR) will increase its 2100 sea level projection derives from a number of post-5AR studies that simulate this at or above 1.2 m and emphasise that global sea level will continue rising for the foreseeable future post-2100. All IPCC reports are freely available online at www.ipcc.ch.

6   Satellite altimetry is a technique that measures the time taken for a radar pulse to travel from the satellite to the ocean surface and back, a measurement from which the sea level can be calculated to an accuracy of less than 1cm. Repeated measurements since the early 1990s have given us maps of the topography of the sea surface as well as the most accurate data ever obtained on the rates of sea level rise.

7   Most of this information comes from a study by Steve Nerem and colleagues (2018, *Proceedings of the National Academy of Sciences of the USA*) who were concerned to demonstrate that sea level variations over the past few decades in regions like the Pacific were plausibly caused largely by a combination of anthropogenic climate change and natural inter-decadal changes.

8   Research shows that where there is adequate 'accommodation space', mangroves can adapt to sea level rise by migrating landwards (Borchert and others, 2018, *Journal of Applied Ecology*).

9   The UK data comes from a 2016 report by the Environment Agency on 'Managing Flood and Coastal Erosion Risks in England'. As for the staggered withdrawal, I learned about this on 9 October 2007 at the residence of the British Ambassador in Tokyo as one of an audience for a talk on climate change by Sir David King, then chief scientific adviser to the British government.

10   Refer to the 2013 article by Toon Haer and others in the journal *Global Environmental Change*.

11   Some 10 per cent of the world's population currently lives in places that are 10m or less above mean sea level (Neumann and others, 2015, *PLoS One*). Population numbers come from the 2016 study by Sylvia Szabo and colleagues in the journal *Sustainability Science*.

12   A fascinating series of articles on this topic appears in the 2015 book, *The Gold Coast Transformed*, edited by Tor Hundloe and others. The specific point that the 'exceedingly fragile' beaches of the Gold Coast would not be able to withstand another succession of events as that which occurred in 1967 is made in the study by Bruno Castelle and others (2008, *Geo-Marine Letters*).

13   Calculations from Keqi Zhang and others (2011, *Climatic Change*).

14   For readers with a legal bent, there is a fascinating article about rethinking contractual impracticability by Myanna Dellinger in the 2016 issue of the *Hastings Law Journal*.

15   The story is told in *The Islanders of the Pacific* by St Johnston (1921). The Marquesan island of Fatu Huku actually did collapse sometime between 1792 and 1820, as explained in Chapter 10.

16   This story comes from Aone van Engelenhoven quoted by Oppenheimer (1998, p. 276).

17   The missionaries on Vanuatu had a hard time coping with the earthquakes, as the many histrionic memoirs written following repatriation to less seismically-active regions attest. 'The house danced, the windows rattled awfully … the heaving must, I think, have continued for nearly five minutes,' wrote Mrs Dr John G. Paton, somewhat ungenerously described by her husband as 'trembling with agitation' during an earthquake on Aniwa Island. It was also Mrs Paton who noted the strange absence on Aniwa of any discernible shaking when the 'awful' earthquake occurred on Tanna Island, 22km away, on 10 January 1878.

18   Not only is Vanuatu a wonderful and safe place to visit (when I was last there, the only women's prison in the country had not had an inmate for years), but it is also a place where you can get uncommonly close to an active volcano. You can visit Tanna Island where at sundown visitors gather along the rim of the Yasur volcano crater. The noise, smell and the astonishing sight of crimson lava bubbling, jets shooting skywards a few tens of metres away from you, all make for a primordial experience that will

leave you in no doubt that our world hosts another reality alongside that with which we are most familiar.

19   It is fascinating to consider that the reason for the formation of the East Antarctic ice sheet has less to do with climate and more with crustal movements. For when Antarctica first became positioned over the South Pole, as it is now, it was still connected by land to South America. Since the coastline of this entire landmass extended from the Pole into the tropics, warm ocean water washed the shores of Antarctica, keeping temperatures higher than you would otherwise expect. But then, the land connection between Antarctica and South America was severed and, for the first time ever, ocean water started to circulate west to east around the entire continent of Antarctica. Warm water could no longer reach its shores. It became 'sealed' by cold water so the temperatures on land started falling and ice began accumulating.

20   The current situation of the Pine Island Glacier and the evidence for its ungrounding were presented by Peter Davies and colleagues (2017, *Journal of Geophysical Research [ES]*) while the discovery of a subsurface volcano in the area was reported by Brice Loose and others (2018, *Nature Communications*).

21   A balanced review by James Hunt and colleagues of the likelihood and consequences of future island–flank collapses was published in 2018 (*Scientific Reports*). Cumbre Vieja is still one to watch. It was first pulled onto the frontburner of geoscience as a result of its eruption on 24 June 1949 when some 200km3 of this volcano abruptly dropped 4m. Then stopped. Much of the subsequent concern arose from the interpretation that this was an 'aborted collapse', which might one day become a full collapse. The current consensus is that this is neither inevitable nor impossible.

22   Every year the Magdalena River contributes 26 per cent of the total freshwater received by the Caribbean Sea, as well as 38 per cent of its total terrestrial sediment load (Restrepo and others, 2016, *Journal of Coastal Research*). This is formidable indeed, bearing in mind that the Mississippi also discharges into the Caribbean.

23   Quote from p. 382 of the 2016 study by Stephen Leslie and Paul Mann (*Earth and Planetary Science Letters*) from which most of the remainder of the information about Magdalena Fan instability is taken.

24   This conversation was in *iTaukei* Bauan. One version of the translation appeared in an article I wrote for *Cosmos* in June 2019 available at https://tinyurl.com/y5g5axkh.

25 This sentiment comes from p. 9 of Barry Cunliffe's erudite 2001 book, *Facing the Ocean*.

26 These quotes are from the first page of Gillis's 2012 book, *The Human Shore*, one of the most thoughtful and entertaining I have read in ages.

27 Note that the original sirens of Greek myth might be either female or male.

28 To keep pace with rising (relative) sea level of 4.1 mm per year, coastal marsh surfaces in Chesapeake Bay would need to build themselves upwards at least at the same rate. Surveys suggest they are not even coming close to this, average wetland surfaces actually decreasing by 1.8 mm per year, something that will inevitably see wetlands converted to open mudflats (Beckett and others, 2016, *PLoS ONE*). Since sea level rise affects tidal range, especially in estuaries like Chesapeake Bay, it is clear that building seawalls in the lower part of the estuary will increase the reach of high tide in its upper parts. Research suggests that by allowing sea level rise to be accommodated (by flooding the land) in lower Chesapeake Bay, its effects in upper Chesapeake Bay will be lessened (Lee and others, 2017, *Journal of Geophysical Research: Oceans*). Hard choices indeed.

Chapter 14: Out of the Shadows: Resisting Land Submergence

1 All quotes from Eric Ash's 2017 book about *The Draining of the Fens*.

2 Quote from p. 2 of the 2006 book, *The Conquest of Nature*, by David Blackbourn.

3 Known in English as the mulberry dike-fish pond complex, this system was developed by the people of the Zhujiang (Pearl) River Delta to optimise its full potential, especially as the region's population grew. It works as follows. The leaves of the mulberry trees are fed to silkworms whose excreta and chrysalises are fed to fish; fish excrement and algae contribute to pond mud which is used as fertiliser for the mulberry trees. Silk and fish sales provide income to farmers (Zhong Gongfu, 1982, *Human Ecology*).

4 While most European colonists in North America created farmland by clearing forests, the Acadians were known as *défricheurs d'eau* (those who cleared the water). Most of the information here comes

from J. Sherman Bleakney's 2004 book, *Sods, Soil and Spades*, while the symbolism in the film *Les Aboiteaux* is discussed in the study by Ronald Rudin (2015, *Canadian Historical Review*).

5   The principal motivation of the English East India Company which was responsible for the land reclamation was to create new land on which food could be produced to feed Bombay's growing population. The Company did not see themselves as uniting islands but rather 'recovering what they thought of as drowned land in the centre of a unitary island' (Riding, 2018, *Journal of Historical Geography*, p. 27). It was only later during the period of the British Raj when efforts were made to aggrandise and promulgate the apparent benefits of Indian colonialism that a narrative involving the uniting of seven islands became popular.

6   A thoughtful study of the potential impacts of climate change on flood risk in Mumbai was published in 2011 by Nicola Ranger and others (*Climatic Change*). It argued that improved drainage could reduce losses associated with a one-in-100-year flood by 70 per cent and that the indirect (economic) effects of flooding could be halved by extending insurance to 100 per cent penetration.

7   Much of this information is from an online article by Yvette Hoitink, used with her permission.

8   Almost all the information in this section comes from the fascinating study by Annet Nieuwhof and colleagues (2019, *Ocean and Coastal Management*). Note that the plural of terp in Dutch is *terpen* but I use the Anglicised word terps in this book. The cereal crops referred to include barley (*Hordeum vulgare*), oats (*Avena sativa*), rye (*Secale cereale*), although it is currently uncertain which of these were being grown at the start of the terp occupation of the Wadden Sea coast.

9   'From the Orinoco to the Amazon, the aqueous fringe of the South American coast has a shallow, muddy, brackish, ochrey sort of composition, which overspreads an almost imperceptible downward slope ... and bears witness to the prodigious volumes of water poured [on it] unceasingly' (Palgrave, 1876, *Dutch Guiana*, pp. 8–9). Some of the earliest missionary endeavours reported the nature of the sodden coastal forested environment less circumspectly; 'Treacherous in the luxuriant loveliness of their vegetation, the forests verified their right to the name ... the land of death. To cut through the dark, deep water in the narrow

coryal might be entrancing ... but oh! the peril of it. Miasma lurks everywhere' (Beach and others, 1900, *Protestant Missions in South America*, p. 51).

10   Most of this information comes from the study by Stéphen Rostain (2010, *Diversity*); the quote is on p. 331.

11   Prominent among these is Robert Van de Noort who favours terp-based coastal communities as a future adaptive strategy in places like the Wadden Sea coast of the Netherlands. His 2013 book, *Climate Change Archaeology*, is a refreshing study that foresees a sustainable future for such populated wetland regions.

12   This extract is from the San Francisco *Alta* of 25 July 1869.

13   And you can discount just about anything written on the subject by Bill Ballinger, especially his daft 1978 book about Nan Madol called *Lost City of Stone* where he suggests that a Greek ship which sailed to India with Alexander the Great took the wrong turn on its way home and reached Pohnpei where the sailors built these artificial islands, the megalithic structures which look like 'American log cabins'. Anything really to avoid the obvious conclusion, which is that the ancestors of modern Pohnpeians built the islands and the structures.

14   The fact we do not yet know where these basalt logs came from or how they were moved to Nan Madol should not be exaggerated. There are many potential sources and the conclusion of Mark McCoy and Steve Athens (2012, *Journal of Pacific Archaeology*) is that the main source may have been mined to exhaustion, which is why it has not been found.

15   The story belongs to the Wati Nyiinyii people of the Nullarbor in southern Australia and is told on p. 104 of my 2018 book.

16   This comes from the thoughtful work of Ian McNiven, in his chapter about the 'sentient sea' in the 2008 *Handbook of Landscape Archaeology*.

17   These ideas come from the research of Serge Cassen, Agnès Baltzer and others, notably in the chapter in the 2011 book on *Submerged Prehistory* edited by Jonathan Benjamin and others in which they state that 'these stone rows acted as a 'cognitive barrier' ... [they] should be considered as a mineral fence that could stop, impede, or filter movement or passage' (p. 100). Somewhat similar is an ancient structure found at Flag Fen (near Peterborough) in the English Fenland which is interpreted as 'a symbolic weir or dam' against the inexorable rise of sea level (Pryor, 2001, *The Flag Fen Basin*).

18 This idea is that of Anja Mansrud, published in the 2017 issue of
    the *International Journal of Nautical Archaeology,* who suggests that
    'fishhooks were animated by certain attributes of the ungulates
    whose bones were used to make them'. The frequent
    representation of elk in Mesolithic rock paintings supports her
    interpretation; some paintings even show boats decorated with
    carved elk heads.
19 This prescient quote comes from p. 61 in the 2003 book
    Environmental Archaeology and the Social Order by John Evans.
20 Alfred Lord Tennyson's poem, *Morte d'Arthur,* was written in
    1842. Several people, including Tennyson, implied that the island
    of Avilion was the (now-submerged) land of Lyonesse (see
    Chapter 8).

Acknowledgements

This book was written while I lived and worked on the traditional lands of the Kabi Kabi people and I wish to acknowledge both the inspiration that they and traditional knowledge-holders the world over have given me, as well as the belief that their wisdom can help us all better understand our place in the world.

At Bloomsbury, I wish to thank Jim Martin, Anna MacDiarmid, Angelique Neumann, Genevieve Nelsson, Julia Mitchell and Emily Kearns.

For various insights and support, I am grateful to Agnès Baltzer, Elizabeth Wayland Barber, Harriot Beazley, Marcus Bussey, Mereoni Camailakeba, Mike Carson, Jen Carter, Serge Cassen, Denise Chow, Amy Clarke, Rita Compatangelo-Soussignan, Axel Creach, Marie-Yvane Daire, Alisha Gauvreau, James Goff, Diane Goodwillie, Ken Greenwood, Christy Haruel, Greg Hill, Yvette Hoitink, Corey Howell, Hsiao-chun Hung, Augustine Kohler, Roselyn Kumar, Loredana Lancini, Frédéric Le Blay, Sepeti Matararaba, Adrian McCallum, Duncan McLaren, Ian McNiven, Daniela Medina Hidalgo, Latha Menon, Elia Nakoro, Meli Nanuku, Sipiriano Nemani, John Runman, Lynette Russell, Damon Salesa, Vasemaca Setariki, Kaliopate Tavola, Frank Thomas and Pila Wilson.

I also want to thank John Gillis and Lynne Kelly for their inspirational writings and Paul Geraghty for always being willing to fill gaps in my knowledge. Sue Svensen and the staff of the Library at the University of the Sunshine Coast have been unfailingly helpful, especially in finding obscure texts. Yet it is to Rosie and Petra that I owe most for helping me complete this book; whether dodging startled kangaroos while climbing Victorian volcanoes, shivering atop wintry windswept Normandy cliffs, or swiping grog at the Fiji Museum, your support never wavered.

Index

Page numbers for maps and illustrations are given in italics.